W0051409

Mathematical Sciences Research Institute
Publications

23

Editors

S.S. Chern
I. Kaplansky
C.C. Moore
I.M. Singer

Mathematical Sciences Research Institute
Publications

G. Baumslag   C.F. Miller III
Editors

Algorithms and Classification in Combinatorial Group Theory

With 13 Figures

Springer-Verlag
New York Berlin Heidelberg London Paris
Tokyo Hong Kong Barcelona Budapest

Gilbert Baumslag
Department of Mathematics
City College (CUNY)
New York, NY 10031
USA

Charles F. Miller III
Department of Mathematics
University of Melbourne
Parkville 3052
Australia

Mathematical Sciences Research Institute
1000 Centennial Drive
Berkeley, CA 94720
USA

The Mathematical Sciences Research Institute wishes to acknowledge support by the National Science Foundation.

Mathematics Subject Classification: 20XXX

Library of Congress Cataloging-in-Publication Data
Algorithms and classification in combinatorial group theory / G.
  Baumslag, C.F. Miller III, editors.
      p. cm. — (Mathematical Sciences Research Institute
  publications : 23)
      Based on lectures presented at the Workshop on Algorithms, Word
  Problems, and Classification in Combinatorial Group Theory, held at
  MSRI, Jan. 1989.
      Includes bibliographical references.
      ISBN-13:978-1-4613-9732-8
      1. Combinatorial group theory—Congresses.   2. Algorithms—
  Congresses.   I. Baumslag, Gilbert.   II. Miller, C. F. (Charles
  F.). 1941– .   III. Workshop on Algorithms, Word Problems, and
  Classification in Combinatorial Group Theory (1989 : Mathematical
  Sciences Research Institute).   IV. Mathematical Sciences Research
  Institute (Berkeley, Calif.).   V. Series.
  QA171.A535 1991   91-5104
  512′.2—dc20

Printed on acid-free paper.

© 1992 Springer-Verlag New York, Inc.
Softcover reprint of the hardcover 1st edition 1992
All rights reserved. This work may not be translated or copied in whole or in part without the written permission of the publisher (Springer-Verlag New York, Inc., 175 Fifth Avenue, New York, NY, 10010, USA), except for brief excerpts in connection with reviews or scholarly analysis. Use in connection with any form of information storage and retrieval, electronic adaptation, computer software, or by similar or dissimilar methodology now known or hereafter developed is forbidden.
The use of general descriptive names, trade names, trademarks, etc., in this publication, even if the former are not especially identified, is not to be taken as a sign that such names, as understood by the Trade Marks and Merchandise Marks Act, may accordingly be used freely by anyone.

Photocomposed copy prepared by the Mathematical Sciences Research Institute using T$_E$X.

9 8 7 6 5 4 3 2 1

ISBN-13:978-1-4613-9732-8   e-ISBN-13:978-1-4613-9730-4
DOI: 10.1007/978-1-4613-9730-4

Preface

In January 1989 a *Workshop on Algorithms, Word Problems and Classification in Combinatorial Group Theory* was held at MSRI. This was part of a year-long program on Geometry and Combinatorial Group Theory organised by Adyan, Brown, Gersten and Stallings. The organisers of the workshop were G. Baumslag, F.B. Cannonito and C.F. Miller III. The papers in this volume are an outgrowth of lectures at this conference.

The first three papers are concerned with decision problems and the next two with finitely presented simple groups. These are followed by two papers dealing with combinatorial geometry and homology. The remaining papers are about automatic groups and related topics.

Some of these papers are, in essence, announcements of new results. The complexity of some of them are such that neither the Editors nor the Reviewers feel that they can take responsibility for vouching for the completeness of the proofs involved.

We wish to thank the staff at MSRI for their help in organising the workshop and this volume.

Gilbert Baumslag
Charles F. Miller III

Algorithms and Classification
in Combinatorial Group Theory

TABLE OF CONTENTS

viii   Contents

Decision Problems for Groups — Survey and Reflections

CHARLES F. MILLER III

1. Introduction

This is a survey of decision problems for groups, that is of algorithms for answering various questions about groups and their elements. The general objective of this area can be formulated as follows:

Objective: To determine the existence and nature of algorithms which decide

- *local properties* – whether or not elements of a group have certain properties or relationships;
- *global properties* – whether or not groups as a whole possess certain properties or relationships.

The groups in question are assumed to be given by finite presentations or in some other explicit manner.

Historically the following three fundamental decision problems formulated by Max Dehn in 1911 have played a central role:

word problem: Let G be a group given by a finite presentation. Does there exist an algorithm to determine of an arbitrary word w in the generators of G whether or not $w =_G 1$?

conjugacy problem: Let G be a group given by a finite presentation. Does there exist an algorithm to determine of an arbitrary pair of words u and v in the generators of G whether or not u and v define conjugate elements of G?

isomorphism problem: Does there exist an algorithm to determine of an arbitrary pair of finite presentations whether or not the groups they present are isomorphic?

In terms of the general objective, the word and conjugacy problems are decision problems about local properties while the isomorphism problem is a decision problem about a global relationship.

Motivation for studying these questions can be found in algebraic topology. For one of the more interesting algebraic invariants of a topological space is its fundamental group. If a connected topological space T is reasonably nice, for instance if T is a finite complex, then its fundamental group $\pi_1(T)$ is finitely presented and a presentation can be found from any reasonable description of T. The word problem for $\pi_1(T)$ then corresponds to the problem of determining whether or not a closed loop in T is contractible. The conjugacy problem for $\pi_1(T)$ corresponds to the problem of determining whether or not two closed loops are freely homotopic (intuitively whether one can be deformed into the other). Since homeomorphic spaces have isomorphic fundamental groups, a solution to the isomorphism problem would give a method for discriminating between spaces (the homeomorphism problem).

Following the development of the theory of algorithms in the 1930's (recursive functions and Turing machines), it was reasonable to expect that Dehn's fundamental problems might be recursively unsolvable. It turns out that not only these problems but a host of local and global decision problems are unsolvable. These developments are discussed in the next two sections. In subsequent sections we consider decision problems restricted to classes of groups enjoying particular algebraic properties, the problem of computing invariants (largely homological) of groups and some measures of computational complexity.

The purpose of this survey is to give some picture of what is known about decision problems in group theory. While a number of references are given for various results, historical matters have been largely neglected. Naturally the choice of material reported on reflects the author's interests and many worthy contributions to the field will unfortunately go without mention. A number of relatively straight forward proofs have been included; usually they are not too difficult, or illustrate the concepts involved or even, occasionally, have a novel aspect. Many concepts and results from mathematical logic, particularly recursive function theory, are explained in an informal manner and occasionally at some length. Hopefully this will make these concepts more accessible for a wide audience.

2. Basic local unsolvability results

A finite presentation π of a group is a piece of notation such as

$$< x_1, \ldots, x_n \mid R_1 = 1, \ldots, R_r = 1 >$$

where the x_i are letters in some fixed alphabet and the R_j are words in the x_i and their inverses x_i^{-1}. The group presented by π, denoted $gp(\pi)$, is the quotient group of the free group on the x_i by the normal closure of the R_j. Usually it is not necessary to distinguish so carefully between a group and its presentation and we often write simply

$$G = < x_1, \ldots, x_n \mid R_1 = 1, \ldots, R_r = 1 >$$

to mean the G is the group defined by the given presentation.

It is convenient to introduce some notation for several decision problems we will consider. Suppose that G is a finitely presented group defined by a presentation as above. Then the *word problem* for G is the decision problem

$$WP(G) = (?w \in G)(w =_G 1).$$

Here the "?" is intended as a sort of quantifier and should be read as "the problem of deciding for an arbitrary word w in G whether or not ...." A closely related problem is the *equality problem*:

$$EqP(G) = (?w_1, w_2 \in G)(w_1 =_G w_2).$$

Of course, $w_1 =_G w_2$ if and only if $w_1 w_2^{-1} =_G 1$ so that an algorithm for solving either of $WP(G)$ or $EqP(G)$ easily yields an algorithm for solving the other. On the other hand, from the viewpoint of computational complexity, these problems are subtly different.

Again using this "?" quantifier, the *conjugacy problem* for G is

$$CP(G) = (?u, v \in G)(\exists x \in G)(x^{-1}ux =_G v).$$

If H is a finitely generated subgroup of G and if H given by say a finite set of words which generate it, then the *generalized word problem* for H in G is the problem of deciding for an arbitrary word w in G whether or not w lies in the subgroup H, that is

$$GWP(H, G) = (?w \in G)(w \in H).$$

When the subgroup H is an arbitrary finitely generated subgroup rather than a fixed one we write simply $GWP(G)$.

On the face of it, each of these algorithmic problems appears to depend on the given presentation. We will show below that the solvability of each of these problems is independent of the finite presentation chosen. It can happen that for a particular finitely presented group each of the above problems is solvable. For instance, if G is a finite group given by a multiplication table presentation, it is easy to describe algorithms for solving $WP(G)$, $CP(G)$ and $GWP(G)$. Similarly, if $F = < x_1, \ldots, x_n \mid >$ is a finitely generated free group $WP(F)$ is solved by freely reducing and $CP(F)$ is solved by cyclically permuting and freely reducing. The $GWP(H, F)$ for finitely generated subgroups H of F is more difficult and its solution is due to Nielsen (see [71]).

Finally, in terms of the "?" notation, the *isomorphism problem* for finitely presented groups is

$$IsoP = (?\pi_1, \pi_2 \text{ finite presentations})(gp(\pi_1) \cong gp(\pi_2)).$$

We assume the reader is familiar with the rudiments of the theory of algorithms and recursive functions. Thus a set of objects is *recursive* if there is an algorithm for deciding membership in the set. A set S of objects is *recursively enumerable* if there is an algorithm for listing all the objects in S. It is easy to see that every recursive set is recursively enumerable. Moreover, a set S is recursive if and only if both S and its complement are recursively enumerable. A diagonal argument can be used to prove the important result that there exists a set which is recursively enumerable but not recursive. This fact is in a sense the source of all undecidability results in mathematics.

Each of the above decision probems is recursively enumerable in the sense that the collection of questions for which the answer is "Yes" is recursively enumerable. For instance, the set of words w of G such that $w =_G 1$ is recursively enumerable. For it is the set of words freely equal to a product of conjugates of the given finite set of defining relations and this set can (in principle) be systematically listed. Thus $WP(G)$ is recursively enumerable. Now $WP(G)$ is recursively solvable (decidable) exactly when the set of words $\{w \in G \mid w =_G 1\}$ is recursive. So $WP(G)$ is recursively solvable if and only if $\{w \in G \mid w \neq_G 1\}$ is recursively enumerable.

Similarly, one can systematically list all true equations between words of G and all true conjugacy equations so that $EqP(G)$ and $CP(G)$ are recursively enumerable. $GWP(H,G)$ is recursively enumerable since one can list the set of all true equations between words of G and words in the generators of H. Finally, if two presentations present isomorphic groups, then one can be obtained from the other by a finite sequence of Tietze transformations. Since the set of presentations obtainable from a given one by a finite sequence of Tietze transformations is recursively enumerable, it follows that $IsoP$ is recursively enumerable.

We recall the notion of Turing reducibility. If A and B are two sets of objects, we write $A \leq_T B$ if an (hypothetical) algorithm to answer questions about membership in B would yield an algorithm to answer questions about A. Thus the decision problem for A is reducible to that for B. One way to make this precise is through the theory of recursive functions. Recursive functions can be defined as the collection of functions obtained from certain base functions (like multiplication and addition) by closing under the usual operations of composition, minimalization and recursion. A function is said to be B-recursive if it is among the functions obtained from the base functions together with the characteristic function for B by closing under the usual operations. Then $A \leq_T B$ is defined to mean that that the characteristic function of A is B-recursive. Of course, if B is already recursive (that is, membership in B is decidable) and if $A \leq_T B$ then A is also recursive.

Now the relation $\leq_T$ is a partial order so we can form the corresponding equivalence relation. Two sets of objects A and B are *Turing equivalent* $A \equiv_T B$ if each is Turing reducible to the other, that is both $A \leq_T B$ and $B \leq_T A$. In terms of this notation there are some obvious relationships among our decision problems:

$$EqP(G) \equiv_T WP(G) \leq_T CP(G)$$

$$WP(G) \equiv_T GWP(1, G) \leq_T GWP(G).$$

We have already observed the first equivalence. Since $w =_G 1$ if and only if w and 1 are conjugate in G it follows that $WP(G) \leq_T CP(G)$. The other assertions are clear.

A *recursive presentation* is a presentation of the form

$$< x_1, \ldots, x_n \mid R_1 = 1, R_2 = 1, \ldots >$$

where $R_1, R_2, \ldots$ is a recursively enumerable set of words. A finitely generated group G is *recursively presented* if it has a recursive presentation. Of course finitely presented groups are recursively presented but the converse is false. The word problem and conjugacy problem are defined for recursively presented groups as before and they are still recursively enumerable problems.

LEMMA 2.1. *Let G be a finitely generated group given by a recursive presentation*

$$G =< x_1, \ldots, x_n \mid R_1 = 1, R_2 = 1, \ldots > .$$

Suppose that H is a finitely generated group with generators $y_1, \ldots, y_m$ and that $\phi : H \to G$ is an injective homomorphism. Then H has a recursive presentation of the form

$$H =< y_1, \ldots, y_n \mid Q_1 = 1, Q_2 = 1, \ldots >$$

where $Q_1, Q_2, \ldots$ is a recursively enumerable set of words in $y_1, \ldots, y_m$. Moreover, $WP(H) \leq_T WP(G)$.

PROOF: Let $F =< y_1, \ldots, y_m \mid >$ be the free group with basis $y_1, \ldots, y_m$. Now we can write $\phi(y_i) = u_i$ $(i = 1, \ldots, m)$ where the u_i are certain words on $x_1, \ldots, x_n$. There is then a unique homomorphism $\psi : F \to G$ such that $\psi(y_i) = u_i$ $(i = 1, \ldots, m)$ and since ϕ is injective we have $H \cong F/\ker \psi$. Now the set of all formal products of the words u_i and their inverses is a recursively enumerable set of words of G. The set of words of G equal to the identity is also recursively enumerable. Hence the intersection of these two sets is a recursively enumerable set of words, and it follows that $ker\ \psi$ is a recursively enumerable set of words on $y_1, \ldots, y_m$. The first claim follows by taking $Q_1, Q_2, \ldots$ to be a recursive enumeration of $ker\ \psi$.

For the second claim, suppose that we have an algorithm A_G to solve the word problem for G. We describe an algorithm to solve the word problem for H as follows: let $w(y_1, \ldots, y_m)$ be an arbitrary word in the generators of H. Since ϕ is injective, $w =_H 1$ if and only if $\phi(w) =_G 1$. Now $\phi(w) = w(u_1, \ldots, u_m)$ so we can apply the algorithm A_G to decide whether or not $w(u_1, \ldots, u_m) =_G 1$. If so, then $w =_H 1$; if not, then $w \neq_H 1$. This algorithm solves the word problem for H. Thus $WP(H) \leq_T WP(G)$ completing the proof.

LEMMA 2.2. *For finitely presented groups (respectively finitely generated, recursively presented groups), the word problem, conjugacy problem and generalized word problem are algebraic invariants. That is, for any two presentations π_1 and π_2 of the same group on a finite set of generators, $WP(\pi_1) \equiv_T WP(\pi_2)$, $CP(\pi_1) \equiv_T CP(\pi_2)$ and $GWP(\pi_1) \equiv_T GWP(\pi_2)$.*

PROOF: The proof is in each case similar to the proof of the second part of the previous lemma except that ϕ is an isomorphism. We omit the details.

The main local unsolvability result is the following:

THEOREM 2.3 (Novikov [87], Boone [21]). *There exists a finitely presented group whose word problem is recursively unsolvable.*

The original proofs of this result proceed along the following lines: start with a Turing machine T whose halting problem is unsolvable. That is, the problem of deciding whether the machine started with an arbitrary tape in a certain state will eventually halt is unsolvable. Constructions of Markov and of Post, associate to such a Turing machine a certain semigroup $S(T)$ whose defining relations mimic the transition rules defining the Turing machine T. They show a code word incorporating a tape and state of T is equal in $S(T)$ to a particular fixed halting word, say q_0, if and only if T halts when started with that tape and state.

Groups $G(T)$ having unsolvable word problem are constructed by in turn mimicking the defining relations of $S(T)$ inside a group. The construction is not so direct as the Markov-Post construction and involves starting with free groups and performing a number of HNN-extensions and/or free products with amalgamation. Nevertheless, there is a direct coding of a tape and state of T as a word w of $G(T)$ so that $w =_{G(T)} 1$ if and only if the machine T halts when started with that tape and state. Since T has an unsolvable halting problem, it follows that $G(T)$ has unsolvable word problem.

A readable account of the Novikov-Boone Theorem along these lines can be found in the textbook by Rotman [92].

In view of the previously noted relationships among our various decision problems, the Novikov-Boone Theorem has the following immediate corollary:

COROLLARY 2.4. *There exists a finitely presented group G such that $WP(G)$, $CP(G)$ and $GWP(G)$ are all recursively unsolvable.*

In contrast to the difficulties encountered for finitely presented groups, it is easy to give examples of finitely generated, recursively presented groups with unsolvable word problem. For example, let $S \subset \mathbf{N}$ be a recursively enumerable set of natural numbers which is not recursive. Define the recursively presented group

$$H_S = < a, b, c, d \mid a^{-i}ba^i = c^{-i}dc^i \; \forall i \in S > .$$

Now H_S can be described as the free product with amalgamation of the free group $< a, b \mid \; >$ and the free group $< c, d \mid \; >$ amalgamating the subgroup (freely) generated by the left hand sides of the indicated equations with the subgroup (freely) generated by the right hand sides. It follows from the normal form theorem for amalgamated free products that $a^{-i}ba^ic^{-i}d^{-1}c^i =_{H_S} 1$ if and only if $i \in S$. Thus $S \leq_T WP(H_S)$ and so $WP(H_S)$ is recursively unsolvable.

Using this observation Higman [54] gave a very different proof of the unsolvability of the word problem. Indeed he proved the following remarkable result:

THEOREM 2.5 (Higman Embedding Theorem). *A finitely generated group H can be embedded in a finitely presented group if and only if H is recursively presented.*

That finitely generated subgroups of finitely presented groups are recursively presented is contained in our first lemma above. The difficult part of this theorem is to show that a recursively presented group can be embedded in a finitely presented group.

The Novikov-Boone Theorem is an easy corollary. For let H_S be the finitely generated, recursively presented group with unsolvable word problem constructed above. By Higman's Embedding Theorem, H_S can be embedded in a finitely presented group, say G_S. Then by an earlier lemma, $WP(H_S) \leq_T WP(G_S)$ and so G_S has unsolvable word problem.

Higman's Embedding Theorem has a number of other remarkable aspects. It provides a complete characterization of the finitely generated subgroups of finitely presented groups - namely they are the recursively presented groups. It also provides a direct connection between a purely algebraic notion and a notion from recursive function theory. Another consequence is the existence of universal finitely presented groups.

COROLLARY 2.6 ([54]). *There exists a universal finitely presented group; that is, there exists a finitely presented group G which contains an isomorphic copy of every finitely presented group.*

To prove this one systematically enumerates all finite presentations on a fixed countable alphabet. The free product of all of these can be embedded in a two generator group which will be recursively presented. This group can then be embedded in a finitely presented group which is the desired universal group.

It is known from recursive function theory that the relation $\equiv_T$ partitions sets into equivalence classes called *degrees of unsolvability*. Those degrees of unsolvability which contain a recursively enumerable set are called *r.e. degrees of unsolvability* and are of particular interest. The r.e. degrees are then partially ordered by $\leq_T$. There is a smallest r.e. degree denoted **0** which consists of the recursive sets and a largest r.e. degree denoted **0'** which is essentially the general halting problem for all Turing machines. However, a particular Turing machine can have a halting problem with degree lying strictly between these two. There are infinitely many r.e. degrees and they have a rich structure; for example, they are dense with respect to the partial order $\leq_T$. In view of this varied collection of r.e. degrees, it is natural to ask which degrees arise from word problems of finitely presented groups. The answer is the following strengthening of the Novikov-Boone Theorem:

THEOREM 2.7 (Fridman [35], Clapham [27], Boone [21]). *Let* **D** *be an r.e.*
degree of unsolvability. Then there is a finitely presented group whose word
problem has degree **D**. *In more detail, there is an explicit, uniform con-*
struction which when applied to a Turing machine T having halting problem
of degree **D** *yields a finitely presented group G(T) such that W P(G(T)) is*
Turing equivalent to the halting problem for T.

The arguments used to prove the Novikov-Boone Theorem already constructed a group $G(T)$ such that the halting problem for T is $\leq_T W P(G(T))$.
The difficulty is in showing the word problem isn't any harder than the halting problem for T. The proofs of this result are technically rather difficult.

It is easy to see that the amalgamated free product of two free groups H_S
described above has $W P(H_S) \equiv_T S$. Clapham's approach to the previous
theorem is to show that Higman's embedding theorem can be made "degree
preserving". More precisely, he shows the following:

THEOREM 2.8 (Clapham [28]). *If H is a finitely generated, recursively*
presented group, then H can be embedded in a finitely presented group G
such that $W P(H) \equiv_T W P(G)$. In particular, a finitely generated group
with solvable word problem can be embedded in a finitely presented group
with solvable word problem.

As we shall see in the next section, the precise control of the word
problem implicit in this result enables one to obtain additional unsolvability
results of a global nature.

Since one always has $W P(G) \leq_T CP(G)$, one can ask whether they are
always the same. They can in fact be of any two appropriate r.e. degrees
of unsolvability. The most general result in this direction is the following:

THEOREM 2.9 (Collins [29]). *Let D_1 and D_2 be r.e. degrees of unsolv-*
ability such that $D_1 \leq_T D_2$. Then there is a finitely presented group G
such that $W P(G)$ has degree D_1 and $CP(G)$ has degree D_2. In partic-
ular, there is a finitely presented group with solvable word problem but
unsolvable conjugacy problem.

We turn now to briefly consider other local decision problems concerning
elements in a group.

The structure of finitely generated abelian groups can be completely
determined from a finite presentation of such a group, and in particular
one can solve the word problem for such groups. Consequently, if G is an
arbitrary finitely presented group one can effectively determine the structure of its abelianization $G/[G, G]$. So for instance, there is an algorithm

to decide whether G is perfect. Moreover, since one can solve the word problem for $G/[G,G]$ it follows that one can decide of a arbitrary word w of G whether or not $w \in [G,G]$.

However, it would seem that any property of elements a finitely presented group which is not determined by the abelianization $G/[G,G]$ will be recursively unrecognizable. The following result show a few common properties of elements are not recognizable.

THEOREM 2.10 (Baumslag, Boone and Neumann [10]). *There is a finitely presented group G such that there is no algorithm to determine whether or not a word in the given generators represents*
   (1) an element of the center of G;
   (2) an element permutable with a given element of G;
   (3) an n-th power, where $n > 1$ is a fixed integer;
   (4) an element whose class of conjugates is finite;
   (5) a commutator;
   (6) an element of finite order > 1.

PROOF: Fix a finitely presented group U having unsolvable word problem. Define G to be the ordinary free product of U with a cyclic group of order 3 and an infinite cyclic group, that is,

$$G = U * < s \mid \ > * < t \mid t^3 = 1 > .$$

We use the commutator notation $[x,y] = x^{-1}y^{-1}xy$. In the following, w is a variable for an arbitrary word in the generators of U.

The center of G is trivial so w lies in the center of G if and only if $w =_U 1$. So there is no algorithm to determine whether an arbitrary word of G lies in the center. This gives the first assertion. Similarly, w is permutable with s if and only if $w =_U 1$ which establishes the second assertion. The element $s^n[t,w]$ is an n-th power if and only if $w =_U 1$ establishing the third assertion. The conjugacy class of w is finite if and only if $w =_U 1$ since if $w \neq_U 1$ the conjugates $s^{-i}ws^i$ would all be distinct. This gives the fourth assertion. For the fifth assertion, note that $[s,t]w$ is a commutator if and only if $w =_U 1$. Finally for the sixth assertion, observe that tw has infinite order if and only if $w \neq_U 1$, while if $w =_U 1$ then tw has order 3. This completes the proof.

In the next section we will present some related unsolvability results concerning the subgroups of a finitely presented group generated by finite sets of elements. These are a sort of mixture between local and global unsolvability results.

3. Basic global unsolvability results

In this section the existence of a finitely presented group with unsovable word problem is applied to obtain a number of global unsolvability results.

Consider the problem of recognizing whether a finitely presented group has a certain property of interest. For example, can one determine from a presentation whether a group is finite? or abelian? It is natural to require that the property to be recognized is *abstract* in the sense that whether a group G enjoys the property is independent of the presentation of G.

An abstract property P of finitely presented groups is *recursively recognizable* if there is an effective method which when applied to an arbitrary finite presentation π determines whether or not $gp(\pi)$ has the property P. More formally, P is *recursively recognizable* if $\{\pi \mid gp(\pi) \in P\}$ is a recursive set of finite presentations.

It turns out that very few interesting properties of groups are recursively recognizable. To formulate the key result we need the following definition.

DEFINITION 3.1: An abstract property P of finitely presented groups is said to be a *Markov property* if there are two finitely presented groups G_+ and G_- such that

(1) G_+ has the property P; and
(2) if G_- is embedded in a finitely presented group H then H does not have property P.

These groups G_+ and G_- will be called the *positive* and *negative witnesses* for the Markov property P respectively.

It should be emphasized that if P is a Markov property then the negative witness does not have the property P, nor is it embedded in any finitely presented group with property P.

For example the property of being finite is a Markov property. For G_+ one can take $< a \mid a^2 = 1 >$ which is a finite group. For G_- one can take the group $< b, c \mid b^{-1}cb = c^2 >$ which is an infinite group and therefore not embedded in any finite group.

Similarly, the property of being abelian is a Markov property. Indeed the two groups chosen as witnesses for the property of being finite will also serve as witnesses for the property of being abelian.

An example of a property which is not a Markov property is the property of being perfect, that is $G/[G, G] \cong 1$. For it is not hard to show (and indeed will follow from the constructions given below) that any finitely presented group can be embedded in a perfect finitely presented group. Hence there can be no negative witness G_- for the property of being perfect.

. An abstract property P of finitely presented groups is *hereditary* if H embedded in G and $G \in P$ imply that $H \in P$, that is, the property P is inherited by finitely presented subgroups. A property of finitely presented groups P is *non-trivial* if it is neither the empty property nor is it enjoyed by all finitely presented groups. Suppose P is a non-trivial, hereditary property of finitely presented groups. Then, since P is non-trivial, there are groups $G_+ \in P$ and $G_- \notin P$. But if G_- is embedded in a finitely presented group H, then $H \notin P$ because P is hereditary. Thus P is a Markov property with witnesses G_+ and G_-. This proves the following:

LEMMA 3.1. *If P is a non-trivial hereditary property of finitely presented groups, then P is a Markov property.*

Another useful observation is the following:

LEMMA 3.2. *If $\emptyset \neq P_1 \subseteq P_2$ are properties of finitely presented groups and if P_2 is a Markov property, then P_1 is also a Markov property.*

For if G_- is a negative witness for P_2 and if $K \in P_1$, then P_1 is a Markov property with positive and negative witnesses K and G_-.

Recall from the previous section that Higman has constructed a universal finitely presented group, say U. If P is a Markov property with positive and negative witnesses G_+ and G_-, then G_- is embedded in U so $U \notin P$. Moreover, if U is embedded in a finitely presented group H then so is G_- and hence $H \notin P$. Thus P is a Markov property with positive and negative witnesses G_+ and U. Hence U is a negative witness for every Markov property.

The main unsolvability result concerning the recognition of properties of finitely presented groups is the following:

THEOREM 3.3 (Adian [2], [3], Rabin [88]). *If P is a Markov property of finitely presented groups, then P is not recursively recognizable.*

Before indicating a proof of this result, we note the following easy corollaries:

COROLLARY 3.4. *The following properties of finitely presented groups are not recursively recognizable:*
  (1) being the trivial group;
  (2) being finite;
  (3) being abelian;

(4) *being nilpotent;*
(5) *being solvable;*
(6) *being free;*
(7) *being torsion-free;*
(8) *being residually finite;*
(9) *having a solvable word problem;*
(10) *being simple;*
(11) *being automatic.*

For each of (1) through (9) is a non-trivial, hereditary property and hence is a Markov property. For (10), it is known (see below) that finitely presented, simple groups have solvable word problem and hence, by the above lemma, being simple is a Markov property. Similarly for (11), automatic groups have solvable word problem and so being automatic is a Markov property.

COROLLARY 3.5. *The isomorphism problem for finitely presented groups is recursively unsolvable.*

For by (1) in the previous corollary there is no algorithm to determine of an arbitrary presentation π whether or not $gp(\pi) \cong 1$.

Proof of the Adian-Rabin Theorem: We are going to give a simple proof of the Adian-Rabin Theorem which is our modification of one given by Gordon [40]. The construction is quite straightforward and variations on the details can be applied to obtain further results. So suppose that P is a Markov property and that G_+ and G_- are witnesses for P. We also have available a finitely presented group U having unsolvable word problem.

Using these three items of initial data, we construct a recursive family of finite presentations $\{\pi_w \mid w \in U\}$ indexed by the words of U so that if $w =_U 1$ then $gp(\pi_w) \cong G_+$ while if $w \neq_U 1$ then G_- is embedded in U. Thus $gp(\pi_w) \in P$ if and only if $w =_U 1$. Since U has unsolvable word problem, it follows that P is not recursively recognizable.

The family $\{\pi_w \mid w \in U\}$ is rather like a collection of buildings constructed from playing cards standing on edge. Such a building can be rather unstable so that if an essential card is removed (corresponding to $w =_U 1$) then the entire structure will collapse. The main technical result needed is the following.

LEMMA 3.6 (Main Technical Lemma). *Let K be a group given by a presentation on a finite or countably infinite set of generators, say*

$$K = < x_1, x_2, \ldots \mid R_1 = 1, R_2 = 1, \ldots > .$$

For any word w in the given generators of K, let L_w be the group with presentation obtained from the given one for K by adding three new generators a, b, c together with defining relations

$$
\begin{align}
(1) && a^{-1}ba &= c^{-1}b^{-1}cbc \\
(2) && a^{-2}b^{-1}aba^2 &= c^{-2}b^{-1}cbc^2 \\
(3) && a^{-3}[w,b]a^3 &= c^{-3}bc^3 \\
(4) && a^{-(3+i)}x_iba^{(3+i)} &= c^{-(3+i)}bc^{(3+i)} \quad i = 1, 2, \ldots
\end{align}
$$

where $[w, b]$ is the commutator of w and b. Then
   (1) if $w \neq_K 1$ then K is embedded in L_w by the inclusion map on generators;
   (2) the normal closure of w in L_w is all of L_w; in particular, if $w =_K 1$ then $L_w \cong 1$, the trivial group;
   (3) L_w is generated by the two elements b and ca^{-1}.

If the given presentation of K is finite, then the specified presentation of L_w is also finite.

PROOF: Suppose first that $w \neq_K 1$. In the free group $< b, c \mid >$ on generators b and c consider the subgroup C generated by b together with the right hand sides of the equations (1) through (4). It is easy to check that the indicated elements are a set of free generators for C since in forming the product of two powers of these elements or their inverses some of the conjugating symbols will remain uncancelled and the middle portions will be unaffected.

   Similarly, in the ordinary free product $K* < a, b \mid >$ of K with the free group on generators a and b consider the subgroup A generated by b together with the left hand sides of the equations (1) through (4). Using the assumption that $w \neq_K 1$ it is again easy to check that the indicated elements are a set of free generators for A.

   Thus assuming $w \neq_K 1$, the indicated presentation for L_w together with the equation identifying the symbol b in each the two factors is the natural presentation for the free product with amalgamation

$$
(K* < a, b \mid >) \quad * \quad < b, c \mid > .
$$
$$
A = C
$$

So if $w \neq_K 1$, then K is embedded in L_w establishing the first claim.

   Now let N_w denote the normal closure of w in L_w. Clearly $[w, b] \in N_w$ so by equation (3), $b \in N_w$. But equations (1) and (2) ensure that a, b, c

are all conjugate and so a, b, c all belong to N_w. Finally, since each of the system of equations (4) can be solved to express x_i in terms of a, b, c, it follows that $x_i \in N_w$ for $i = 1, 2, \ldots$. Thus each of the generators of L_w belongs to N_w and so $L_w = N_w$. This verifies the second assertion.

Finally, let M be the subgroup of L_w generated by b and ca^{-1}. Equation (1) can be rewritten as $b(ca^{-1})b(ca^{-1})^{-1}b^{-1} = c$ so that $c \in M$. But then from $ca^{-1} \in M$ it follows that $a \in M$. Finally from the system of equations (4) which can be solved for the x_i in terms of a, b, c it follows that $x_i \in M$ for $i = 1, 2, \ldots$ and so $M = L_w$. (For later use we note that neither equation (2) nor equation (3) was used in the proof of the final assertion). This completes the proof of the lemma.

Using this technical lemma it is easy to complete the proof of the Adian-Rabin Theorem. We are given the three finitely presented groups U, G_+ and G_- which can be assumed presented on disjoint alphabets as follows:

$$U = < y_1, \ldots, y_k \mid Q_1 = 1, \ldots, Q_q = 1 >$$

$$G_- = < u_1, \ldots, u_m \mid S_1 = 1, \ldots, S_s = 1 >$$

$$G_+ = < v_1, \ldots, v_n \mid T_1 = 1, \ldots, T_t = 1 >$$

Let $K = U * G_-$ the ordinary free product of U and G_- presented as the union of the presentations of its factors. Since U has unsolvable word problem, K also has unsolvable word problem. Also both U and G_- are embedded in K by the inclusion map on generators. For any word w in the generators of U (these are also generators of K) form the presentation L_w as in the Main Technical Lemma. Finally we form the ordinary free product $L_w * G_+$.

A presentation π_w for these groups $L_w * G_+$ can be obtained by simply writing down all of the above generators together with all of the above defining equations. Such a presentation is defined for any word w in U whether or not $w \neq_U 1$. But it follows from the lemma that if $w \neq_U 1$ then the group G_- is embedded in $gp(\pi_w) = L_w * G_+$ and so $gp(\pi_w) \notin P$ by the definition of a Markov property. On the other hand, if $w =_U 1$ then by the lemma $L_w \cong 1$ and so $gp(\pi_w) \cong G_+$ and hence $gp(\pi_w) \in P$.

Thus we have shown that the recursive collection of presentations

$$\{\pi_w \mid w \text{ a word in } U\}$$

has the property that $gp(\pi_w) \in P$ if and only if $w =_U 1$. Since U has unsolvable word problem, it follows that P is not recursively recognizable. This completes the proof of the Adian-Rabin Theorem.

The Main Technical Lemma (and minor variations on it) can be used to establish a number other results. Here are some well-known results which follow easily:

COROLLARY 3.7 (Higman, Neumann, and Neumann [55]). *Every countable group K can be embedded in a two generator group L. If K can be presented by n defining relations, then L can be chosen to have n defining relations.*

PROOF: Since K is countable it can be presented as in the statement of the Main Technical Lemma. Form L as in the lemma except omit the two defining relations (2) and (3). Only equation (3) involved the parameter w and neither equation (2) nor equation (3) were used in the proof that L is two generator. Then K is embedded in L which is a two generator group with generators b and ca^{-1}. This proves the first assertion. (At the expense of considering cases, one could instead use the lemma as stated by choosing a fixed $w \neq_K 1$.)

Equation (1) of the lemma defines c in terms of these two generators. Then using a as an abbreviation for $(ca^{-1})^{-1}c$, the system of equations (4) define the x_i in terms of the given generators. Hence equation (1) and all of the equations of (4) can be eliminated, leaving only the relations R_j rewritten in terms of the generators b and ca^{-1}. This completes the proof.

Combining the proof of this corollary with the Higman Embedding Theorem we obtain the following result which is frequently useful.

COROLLARY 3.8. *If the group K can be presented by a recursive set of generators subject to a recursively enumerable set of defining relations, then K can be embedded in a two generator, finitely presented group L. Under this embedding the given generators of K are represented by a recursive set of words in the generators of L.*

PROOF: As in the proof of the previous corollary, the group K can be embedded in a two generator recursively presented group, say L_1. By the Higman Embedding Theorem this recursively presented group L_1 can be embedded in a finitely presented group, say K_1. Now again applying the previous corollary, K_1 can be embedded in a two generator finitely presented group L as desired. The final assertion follows from the explicit nature of the embedding of K into the two generator group L_1.

Mixing these constructions with results from recursive function theory, we obtain the following result which has a number of applications, for instance to the study of algebraically closed groups.

COROLLARY 3.9 (Miller [78]). *There exists a finitely presented group G with unsolvable word problem such that every non-trivial quotient group of G also has unsolvable word problem.*

PROOF: A pair of disjoint sets of natural numbers P and Q is said to be *recursively inseparable* if there is no recursive set R such that $P \subseteq R$ and $Q \cap R = \emptyset$. The result we need from recursive function theory is that there exists a disjoint pair of recursively enumerable sets P and Q which are recursively inseparable. We may suppose that these sets are chosen so that $0 \in P$ and $1 \in Q$. Define K_0 to be the group presented by

$$K_0 = < e_0, e_1, e_2, e_3, \ldots \mid e_0 = e_i \ \forall i \in P, \quad e_1 = e_j \ \forall j \in Q > .$$

Since P and Q are recursively enumerable, the previous corollary implies that K_0 can be embedded in a two generator, finitely presented group which we will denote by K. We continue to use the symbol e_k as an abbreviation for the word in K which is the image of image of e_k under the embedding. Now apply the Main Technical Lemma above to K and the word $e_0 e_1^{-1}$ to obtain the finitely presented group $G = L_{e_0 e_1^{-1}}$. Since P and Q were disjoint, $e_0 \neq_K e_1$ or equivalently $e_0 e_1^{-1} \neq_K 1$.

Now suppose that H is a non-trivial quotient group of G. We view H as being presented on the same set of generation symbols as G. Since H is non-trivial, by the second assertion of the Main Technical Lemma $e_0 \neq_H e_1$. Put $R = \{i \mid e_0 =_H e_i\}$. Since H is a quotient group of G it follows that $P \subseteq R$. But since $e_0 \neq_H e_1$ it follows that $Q \cap R = \emptyset$. Because P and Q are recursively inseparable, R is not recursive. Now the $\{e_k\}$ are a recursive set of words in the generators of H so if H had a solvable word problem then R would be recursive. Hence H must have an unsolvable word problem. This completes the proof.

COROLLARY 3.10 (P. Hall [52], Goryushkin [42], Schupp [97]). *Every countable group K can be embedded in a 2-generator simple group.*

PROOF: If $K \cong 1$ the result is clear. Suppose $K \not\cong 1$ and let $x_1, x_2, \ldots$ be a list of the non-trivial elements of K and take K to be presented on these generators. Form the two generator group L as in the Main Technical Lemma except that the equations (3) and (4) are to be replaced by the two systems of equations

(3) $\qquad a^{-(3+2i)}[x_i, b]a^{(3+2i)} \ = \ c^{-(3+2i)}dc^{(3+2i)} \quad i = 1, 2, \ldots$

(4) $\qquad a^{-(4+2i)}x_iba^{(4+2i)} \ = \ c^{-(4+2i)}dc^{(4+2i)} \quad i = 1, 2, \ldots .$

Note that the commutator $[x_i, b]$ has infinite order since each $x_i \neq_K 1$ and that K is embedded in L.

Using Zorn's lemma, choose a normal N subgroup of L maximal with respect to the property $K \cap N = 1$. The normal closure in L of any x_i contains $[x_i, b]$ and hence d and is thus clearly all of L. It follows that L/N is a two generator simple group containing an isomorphic copy of K. This completes the proof.

The Adian-Rabin Theorem asserts that a large number of properties of finitely presented groups are not recursive. Recursive function theory provides a number of methods for classifying the difficulty of decision problems so it is natural to ask how difficult are various properties of groups to recognize?

Many properties of interest are recursively enumerable. For instance, in the case of the property "being trivial", the collection of all finite presentations of the trivial group is recursively enumerable. To see this one simply observes that each such presentation can be obtained from the obvious one $< x \mid x = 1 >$ by a finite sequence of Tietze transformations and the set of all Tietze transformations of any finite presentation is recursively enumerable.

Similarly, the property of "being abelian" is recursively enumerable. For we know a canonical form in which to present a finitely presented abelian group; that is, these nice presentations are recognizable and every finitely presented abelian group has such a presentation. Since every other presentation can be obtained from a canonical one by Tietze transformations, it follows that "being abelian" is a recursively enumerable property. Similarly being finite, nilpotent, polycyclic and free are all recursively enumerable properties.

An appropriate method for trying to classify familiar group theoretic properties is to try to locate them in the *arithmetic hierarchy*. Recall that a property or relation P is recursively enumerable (r.e.) if and only if it can be expressed in the form $\exists x R$ where R is a recursive relation involving an additional variable. (A coding device can be used to reduce the apparently more general form $\exists x_1 \ldots \exists x_n R_n$ to a single existential quantifier.) The set of relations expressible in this form is denoted Σ_1^0. The superscript "0" here is to distinguish the arithmetic hierarchy we are going to describe from others. The set of relations expressible in the form $\forall x R$ where R is a recursive relation is denoted Π_1^0. Now the complement of an r.e. relation lies in Π_1^0. For if P has the form $\exists x R$ where R is a recursive relation then $\neg P$ has the form $\forall x \neg R$. Since $\neg R$ is also recursive, it follows that $\neg P$ is in Π_1^0. Recall that a relation P is recursive if and only if both P and $\neg P$ are recursively enumerable. Thus P is recursive if and only if $P \in \Sigma_1^0 \cap \Pi_1^0$.

More generally, a relation P is said to be in Σ_n^0 if it is expressible in the form $\exists \forall \ldots R$ where there are n alternations of quantifiers. (As mentioned before, adjacent quantifiers of the same type can be collapsed into a single quantifier; only the number of alternations matters.) Thus for example $P \in \Sigma_3^0$ means that $P \leftrightarrow \exists x_1 \forall x_2 \exists x_3 R$ for some recursive relation R. Similarly P is said to be in Π_n^0 if it is expressible in the form $\forall \exists \ldots R$ where there are n alternations of quantifiers. Thus $P \in \Pi_2^0$ means that $P \leftrightarrow \forall x_1 \exists x_2 R$ for some recursive relation R.

We need a few fundamental facts about this arithmetic hierarchy. First, if P is a recursive relation then P is in Σ_n^0 and in Π_n^0 for all $n \geq 0$. If P is in Σ_m^0 or in Π_m^0, then P is in Σ_n^0 and in Π_n^0 for all $n > m$. In particular $\Sigma_m^0 \cup \Pi_m^0 \subseteq \Sigma_{m+1}^0 \cap \Pi_{m+1}^0$. Also P is in Σ_n^0 if and only if its complement $\neg P$ is in Π_n^0.

Finally there is the result called the Arithmetical Hierarchy Theorem due to Kleene which asserts these classes are increasingly difficult: for each $n \geq 1$ there is a unary Σ_n^0 relation P which is not Π_n^0 and hence not Σ_n^0 or Π_m^0 for any $m < n$. Hence also $\neg P$ is Π_n^0 but not Σ_n^0 or Π_m^0 for any $m < n$. This is of course a generalization of the existence of r.e. but non-recursive sets. In addition, one can show that each Σ_n^0 contains certain "most difficult" relations not in Π_n^0 which are said to be Σ_n^0-*complete* (and similarly there are Π_n^0-complete relations).

While many properties of groups may be recursively enumerable, it turns out that the property of "having a solvable word problem" is very far from recursively enumerable. Recall that there are constructions which yield groups with word problem equivalent to the halting problem for given Turing machines (Theorem 1.7). Combining this with some results from recursive function theory, Boone and Rogers have shown the following:

THEOREM 3.11 (Boone and Rogers [24]). *For finitely presented groups, the property of having a solvable word problem is Σ_3^0-complete.*

This result has a number of striking consequences about the general enterprise of solving the word problem for finitely presented groups, some of which are the following:

COROLLARY 3.12 ([24]). *There is no recursive enumeration*

$$G_0, G_1, G_2, \ldots$$

of all finitely presented groups with solvable word problem.

For if there were such an enumeration, then all finite presentations of groups having solvable word problem would be recursively enumerable, that is, Σ_1^0. But by the theorem this is impossible since having a solvable word problem is Σ_3^0-complete.

Similarly combining the Theorem 1.7 with a construction applied to Turing machines Boone and Rogers establish the following:

THEOREM 3.13 ([24]). *There is no uniform partial algorithm which solves the word problem in all finitely presented groups with solvable word problem.*

Combining this last corollary with an enumeration of homomorphisms one can show the following result.

COROLLARY 3.14 (Miller [77]). *There is no universal solvable word problem group. That is, if G is a finitely presented group which contains an isomorphic copy of every finitely presented group with solvable word problem, then G itself must have unsolvable word problem.*

More generally, one can ask what is the level in the arithmetic hierarchy of any Markov property of interest. If it is r.e. then the Adian-Rabin result can be used to show it is Σ_1^0-complete. Upper bounds can be found for those not apparently r.e., but the status of several properties remains unresolved. The table below indicates some of the unresolved issues.

Markov property	recursion theoretic status
being trivial	r.e. (Σ_1^0-complete)
being finite	r.e.
being abelain	r.e.
being nilpotent	r.e.
being polycyclic	r.e.
being solvable	? ($\leq \Sigma_3^0$)
being free	r.e.
being torsion-free	? ($\leq \Pi_2^0$)
being residually finite	? ($\leq \Pi_2^0$)
having solvable word problem	Σ_3^0-complete
being simple	? ($\leq \Pi_2^0$)
being automatic	r.e.

We turn now to some unsolvability results concerning the recognition of certain properties of the subgroups of a finitely presented group generated by finite sets of elements. The following is a general result of this type motivated by the Adian-Rabin Theorem.

THEOREM 3.15 (Baumslag, Boone and Neumann [10]). *Let P be an algebriac property of groups and assume that (i) there is a finitely presented group that has P and that (ii) there is an integer n such that no free group F_r of rank $r \geq n$ has P. Then there is a finitely presented group G_P such that there is no algorithm to determine whether or not the subgroup generated by an arbitrary finite set of words in the given generators of G_P has the property P.*

PROOF: Fix a finitely presented group U having unsolvable word problem. Let G_+ be a finitely presented group which has property P. By hypothesis, there is an integer n such that the free groups F_r of rank $r \geq n$ do not have P. We may assume that G_+ is generated by at least n elements, say by $t_1, \ldots, t_r (r \geq n)$. Now form the ordinary free product G_P of U with G_+ and a free group of rank two, that is,

$$G_P = U * G_+ * < a, b \mid >.$$

We use w as a variable for words in the generators of U. If $w \neq_U 1$ then in G_P the elements

$$a^{-1}b^{-1}[a,w]ba, \ldots, a^{-r}b^{-1}[a,w]ba^r$$

freely generate a free subgroup of rank r. Hence also the elements

$$t_i a^{-i} b^{-1} [a,w] ba^i \quad \text{for } i = 1, \ldots, r$$

freely generate a free subgroup of rank r which does not have P. On the other hand, if $w =_U 1$ then $t_i a^{-i} b^{-1}[a,w]ba^i =_{G_P} t_i$ so in this case these elements just generate G_+ which has P. That is, the elements $t_i a^{-i} b^{-1}[a,w]ba^i$ for $i = 1, \ldots, r$ generate a subgroup having the property P if and only if $w =_U 1$. Since U has unsolvable word problem, the result follows.

COROLLARY 3.16 ([10]). *There are finitely presented groups G (depending on the property considered) such that there is no algorithm to determine whether or not the subgroup generated by an arbitrary finite set of words of G is*
  (1) trivial;
  (2) finite;
  (3) free;
  (4) locally free;
  (5) cyclic;
  (6) abelian;

(7) nilpotent;

(8) soluble;

(9) simple;

(10) directly decomposable;

(11) freely indecomposable;

(12) a group with solvable word problem.

Using similar constructions, Baumslag, Boone and Neumann further show the following which is similar to the above corollary but not an immediate consequence of the previous theorem:

THEOREM 3.17 ([10]). *There are finitely presented groups G (depending on the property considered) such that there is no algorithm to determine whether or not the subgroup generated by an arbitrary finite set of words of G is*

(1) *a finitely related subgroup;*

(2) *a subgroup of finite index;*

(3) *a normal subgroup;*

(4) *a subgroup with finitely many conjugates.*

4. Decision problems and constructions

In this section we consider the basic decision problems for finitely presented groups which are built from more elementary groups by such operations as direct products, extensions, free products, amalagamated free products, and HNN-extensions.

The basic decision problems are well behaved with respect to (ordinary) free products. Suppose that A and B are finitely presented groups with solvable word problem (respectively, solvable conjugacy problem). Then their free product $A * B$ has solvable word problem (respectively, solvable conjugacy problem) by stantard results on normal forms and conjugacy in free products. Mihailova [76] has shown the more difficult result that if A and B have solvable generalized word problem, then $A * B$ also has solvable generalized word problem. Concerning the isomorphism problem, the Grushko-Neumann theorem easily implies the following: If C is a recursive class of freely indecomposable finitely presented groups and if there is an algorithm to decide the isomorphism problem for groups in C, then there is an algorithm to decide the isomorphism problem for the class of all free products of finitely many groups in C.

One might have thought the direct product was a rather benign construction. For example it is clear that if A and B are finitely presented groups with solvable word problem (respectively, solvable conjugacy problem) then their direct product $A \times B$ has solvable word problem (respectively, solvable conjugacy problem). However, Mihailova [75] has shown that solvability of the generalized word problem is not preserved by direct products. The proof is based on the following lemma.

LEMMA 4.1. *Let M be any group with a given set of generators $\{s_1, \ldots, s_n\}$ having quotient group H with presentation*

$$H = < s_1, \ldots, s_n \mid R_1 = 1, \ldots, R_m = 1 >$$

on the images of the given generators of M. Let $G = M \times M$ be the direct product of two copies of M and let L_H be the subgroup of G generated by the elements

$$(s_1, s_1), (s_2, s_2), \ldots, (s_n, s_n)$$
$$(R_1, 1), (R_2, 1), \ldots, (R_m, 1).$$

Then for any pair of word u and v in the given generators,

$$(u, v) \in L_H \text{ if and only if } u =_H v.$$

Before beginning the proof, we note that the subgroup L_H is just the pull-back or fibre-product of two copies of the quotient mapping from M onto H.

PROOF: Clearly if $(u, v) \in L_H$ then $u =_H v$ since this is true for each of the generators of L_H and H is a quotient group of M.

For the converse, first suppose $w =_H 1$. Then

$$w =_M \prod_{k=1}^{r} X_k(s_i)^{-1} R_{j_k}^{\epsilon_k} X_k(s_i)$$

for suitable words X_k in the given generators. But then in $G = M \times M$ we have

$$(w, 1) =_G \prod_{k=1}^{r} X_k((s_i, s_i))^{-1} (R_{j_k}^{\epsilon_k}, 1) X_k((s_i, s_i)) \in L_H$$

as desired. More generally suppose $u =_H v$. Then $uv^{-1} =_H 1$ and so $(uv^{-1}, 1) \in L_H$. Since L_H contains the diagonal of $G = M \times M$, it contains (v, v) and hence we also have $(u, v) \in L_H$. This completes the proof.

THEOREM 4.2. *Let M be a finitely presented group having a quotient group H with unsolvable word problem. Then the group $G = M \times M$ has a finitely generated subgroup L_H such that the generalized word problem for L_H in G is recursively unsolvable.*

PROOF: According to the previous lemma using the same notation, $(w, 1) \in L_H$ if and only if $w =_H 1$. Since the word problem for H is unsolvable, the problem of deciding membership in L_H is recursively unsolvable. This proves the theorem.

COROLLARY 4.3 (Mihailova [75]). *Let F be a finitely generated free group of rank at least 2. Then the group $G = F \times F$ has a finitely generated subgroup L such that the generalized word problem for L in G is recursively unsolvable.*

The above lemma can also be used to obtain a number of other unsolvability results concerning the direct product of two free groups. The following result was proved for free groups of rank at least nine by Miller [77] and later improved to two generators by Schupp [97]

THEOREM 4.4 (Miller [77]). *Let F be a free group of rank at least 2 and let $G = F \times F$. The the problem to determine of an arbitrary finite set of words whether or not they generate G is recursively unsolvable.*

PROOF: To see this we apply the proof of the Adian-Rabin Theorem in the case of the Markov property "being trivial" so that $G_+ = 1$. In this case one obtains a recursive family of presentations $\{\pi_w \mid w \in U\}$ indexed by words in a group U with unsolvable word problem such that $\pi_w \cong 1$ if and only if $w =_U 1$. Moreover, if we suppose F has free basis $\{s_1, \ldots, s_n\}$, then each π_w can be written as a presentation on the same generating symbols. For by the Main Technical Lemma the presentations could be given on two generators, say s_1 and s_2. In case $n > 2$ the same group can then also be presented by adding additional generators $\{s_3, \ldots, s_n\}$ and additional defining relations $s_3 = 1, \ldots, s_n = 1$.

Now in the above lemma take $M = F$ and let L_w be the subgroup generated by the indicated elements using the presentation π_w as H. Let $H_w = gp(\pi_w)$. Then by the lemma and properties of the π_w we have

$$
\begin{aligned}
L_w = G \quad &\leftrightarrow \quad \forall u, v \in F((u, v) \in L_w) \\
&\leftrightarrow \quad \forall u, v \in F(u =_{H_w} v) \\
&\leftrightarrow \quad H_w \cong 1 \\
&\leftrightarrow \quad w =_U 1.
\end{aligned}
$$

Since U has unsolvable word problem, this proves the result.

Continuing to use the notation of the previous proof, let N_w be the kernel of the natural homomorphism from F onto H_w. Now by the above lemma $(y, 1) \in L_w$ if and only if $y \in N_w$. Thus the intersection of L_w with the first factor $F \times \{1\}$ of G is N_w (or more precisely $N_w \times \{1\}$). If $w \neq_U 1$ then from the proof of the Adian-Rabin Theorem, we know U is embedded ·in H_w and so in particular H_w is infinite. So if $w \neq_U 1$ then N_w is not finitely generated.

Now in $F \times F$ one can check that the centralizer of any element is finitely generated. On the other hand the centralizer of an element $(1, z) \in L_w$ in L_w will have the form $N_w \times \mathbf{Z}$. So if $w \neq_U 1$ then the centralizer of an element of L_w need not be finitely generated. Consequently, if $w \neq_U 1$ then L_w and $G = F \times F$ are not isomorphic. This proves the following:

THEOREM 4.5. *Let F be a free group of rank at least 2 and let $G = F \times F$. Then the problem to determine of an arbitrary finite set of words whether or not they generate a subgroup isomorphic to G is recursively unsolvable. Hence the isomorphism problem for subgroups of G given by finite sets of generators is recursively unsolvable.*

Variations on the above arguments can be used to show the following:

THEOREM 4.6 (Miller [77]). *Let F be a free group of rank at least two. Put $G = F \times F$. Then G has a finitely generated subgroup L such that L has an unsolvable conjugacy problem. Moreover, the generalized word problem for L in G is unsolvable.*

Notice that the group $F \times F$ is residually free as are all of its subgroups. However, the finitely generated subgroups L constructed in the above are not finitely presented (see Baumslag and Roseblade [19]).

Recall that a property P of groups is a *poly-property* if, whenever N and G/N have P, so has G. Here N is a normal subgroup of G.

LEMMA 4.7. *The following are poly-properties*
  (1) *being finitely generated*
  (2) *having a finite presentation*
  (3) *satisfying the maximum condition for subgroups*
  (4) *being finitely presented and having a solvable word problem.*

That the first three are poly-properties is shown in Hall [51] and that the last is a poly-property is easily verified. Note that it is not enough in the last item to assume finite generation because the extension might not then be even recursively presented (see [12]).

Since $F \times F$ where F is free can have unsolvable generalized word problem, the property "having solvable generalized word problem" is not a poly-property. Neither is "having solvable conjugacy problem". Indeed the following result shows groups with a surprisingly elementary structure can have an unsolvable conjugacy problem:

THEOREM 4.8 (Miller [77]). *There is a finitely presented group G with the following properties:*
  (1) G is the split extension of one finitely generated free group by another, that is, there is a split short exact sequence of groups $1 \to F \to G \to T \to 1$ where F and T are finitely generated free groups.
  (2) G is residually finite;
  (3) G has solvable word problem;
  (4) G has unsolvable conjugacy problem.

Also, G is an HNN-extension of the finitely generated free group F with a finite number of stable letters (the generators of T) acting as automorphisms of F. The ordinary free product $G * T_0$ of G with a free group $T_0 \cong T$ has the structure of an amalgamated free product of two finitely generated free groups with finitely generated amalgamation.

PROOF: (Sketch) Suppose that

$$U = < s_1, \ldots, s_n \mid R_1 = 1, \ldots, R_m = 1 >$$

is a finitely presented group with unsolvable word problem. Let $F = < q, s_1, \ldots, s_n \mid >$ be the free group of rank $n + 1$ on the listed generators. The idea is to construct a group G in which the equations of U are mimicked by conjugations of certain words in F. Of course U can not be embedded in G since G is to have solvable word problem. G is defined as follows:
  Generators: $q, s_1, \ldots, s_n, d_1, \ldots, d_n, t_1, \ldots, t_m$.
  Defining relations:

$$\begin{aligned} t_i^{-1} q t_i &= q R_i \\ t_i^{-1} s_j t_i &= s_j \end{aligned} \left. \right\} \begin{aligned} 1 \le i \le m \\ 1 \le j \le n \end{aligned}$$

$$\begin{aligned} d_k^{-1} q d_k &= s_k^{-1} q s_k \\ d_k^{-1} s_j d_k &= s_j \end{aligned} \left. \right\} \begin{aligned} 1 \le k \le n \\ 1 \le j \le n \end{aligned}$$

For each d_k and t_i the right hand sides of the above defining relations generate the free group F and so by a theorem of Nielsen [71] they freely generate F. Hence these relations define an action of each d_k and t_i as an automorphism of F. If we denote by T the free group with basis $\{d_1, \ldots, d_n, t_1, \ldots, t_n\}$, then the quotient of G by its normal subgoup F is clearly isomorphic to T. A group with this sort of structure must be

residually finite (see [77]). The assertions about HNN-extensions and amalgamated free products follow from general facts about those constructions.

To see that G has unsolvable conjugacy problem, one shows the follow-
ing: if w is any word on $\{s_1, \ldots, s_n\}$, then qw is conjugate in G to q if and
only if $w =_U 1$. Since the word problem for U is unsolvable, it follows that
the conjugacy problem for G is unsolvable.

In one direction this claim is easy. For observe that U is the homomorphic image of G obtained by mapping each s_i in G to the corresponding s_i
in U and mapping all other generators to the identity 1_U. Denote this homomorphism by ϕ. Then if $y^{-1}qwy =_G q$ it follows that $\phi(y)^{-1}w\phi(y) =_U 1$
and hence $w =_U 1$. For the converse, suppose $w =_U 1$. Then

$$w =_U \prod_{k-1}^{r} X_k(s_i)^{-1} R_{j_k}^{\epsilon_k} X_k(s_i)$$

for suitable words X_k in the given generators. As an example of a conjugation in G, consider the following:

$$
\begin{aligned}
(d_1 d_2 t_i^\epsilon d_2^{-1} d_1^{-1})^{-1} q d_1 d_2 t_i^\epsilon d_2^{-1} d_1^{-1} &=_G d_1 d_2 t_i^{-\epsilon} d_2^{-1} d_1^{-1} q d_1 d_2 t_i^\epsilon d_2^{-1} d_1^{-1} \\
&=_G d_1 d_2 t_i^{-\epsilon} d_2^{-1} s_1^{-1} q s_1 d_2 t_i^\epsilon d_2^{-1} d_1^{-1} \\
&=_G d_1 d_2 t_i^{-\epsilon} s_1^{-1} s_2^{-1} q s_2 s_1 t_i^\epsilon d_2^{-1} d_1^{-1} \\
&=_G d_1 d_2 s_1^{-1} s_2^{-1} q R_i^\epsilon s_2 s_1 d_2^{-1} d_1^{-1} \\
&=_G d_1 s_1^{-1} q s_2^{-1} R_i^\epsilon s_2 s_1 d_1^{-1} \\
&=_G q s_1^{-1} s_2^{-1} R_i^\epsilon s_2 s_1
\end{aligned}
$$

Generalizing this calculation, one can find a word Y on $\{d_1, \ldots, d_n, t_1, \ldots, t_n\}$
determined by the representation of w as a product of conjugates of the
defining relations of U such that $Y^{-1}qY =_G qw$ which is the desired result
(see [77] for more details of the calculation). This completes our sketch of
the proof.

The construction of the previous theorem can be combined with the
construction used in proving the Adian-Rabin Theorem to show that the
isomorphism problem for groups with such a very elementary structure is
unsolvable. The details are somewhat more difficult.

THEOREM 4.9 (Miller [77]). *Let U be a group with unsolvable word prob-
lem. Then there is a recursive class of finite presentations $\Omega = \{\pi_w \mid w \in U\}$ indexed by words of U such that*
(1) each $gp(\pi_w)$ is residually finite;

(2) each $gp(\pi_w)$ is the split extension of one finitely generated free group by another;

(3) the word problem for each of the groups $gp(\pi_w)$ is solvable by a uniform method;

(4) $gp(\pi_w) \cong gp(\pi_1)$ if and only if $w =_U 1$.

In particular, the isomorphism problem for Ω is unsolvable. Hence the isomorphism problem for finitely presented, residually finite groups is unsolvable.

The foregoing results show, among other things, that "having solvable generalized word problem" and having "solvable conjugacy problem" are not poly-properties. It is natural to ask whether these properties are at least preserved under finite extensions. For the generalized word problem the answer is not difficult.

LEMMA 4.10 ([11]). *If G is a finite extension of the finitely generated group H having solvable generalized word problem, then G has solvable generalized word problem.*

In contrast to this, the conjugacy problem for a group G and for a subgroup of finite index in G can be quite different as the following result shows:

THEOREM 4.11 ([30]).

(1) *There is a finitely presented group G_1 with unsolvable conjugacy problem that has a subgroup M of index 2 which has solvable conjugacy problem.*

(2) *There is a finitely presented group G_2 with solvable conjugacy problem that has a subgroup L of index 2 which has unsolvable conjugacy problem.*

An example of the first type was given by Gorjaga and Kirkinskii [41], while examples of both of these phenomena were given by Collins and Miller [30]. The proofs are somewhat technical.

A further aspect of the above results is the following: denote by Ψ_0 the set of finitely generated free groups. Then denote by Ψ_1 the collection of groups formed from groups in Ψ_0 by either free product with finitely generated amalgamation or HNN-extension with finitely many stable letters and finitely generated associated subgroups. Similarly form Ψ_2 by applying these constructions to groups in Ψ_1. Note that all of these groups are finitely presented.

By a theorem of Nielsen [71], finitely generated free groups have solvable generalized word problem so each of the groups in Ψ_1 has a solvable word problem. However, the foregoing results show that the generalized word problem, the conjugacy problem and the isomorphism problem can be recursively unsolvable for groups in Ψ_1. Since the generalized word problem is unsolvable for a group in Ψ_1, it follows that the word problem for a suitable group in Ψ_2 is unsolvable.

For example, if F is free of rank at least two and L is a finitely generated subgroup such that the generalized word problem for L in $F \times F$ is unsolvable, one can form the HNN-extension

$$G =< F \times F, t \mid t^{-1}xt = x, x \in L > .$$

Then for any word $y \in F \times F$ we have $t^{-1}yt = y$ if and only if $y \in L$. Since membership in L is not decidable, the word problem for G is unsolvable. Of course $G \in \Psi_2$ and G is finitely presented.

Using the Mayer-Viettoris sequences for (co)homology of free products with amalgamation and for HNN-extensions, it is easy to check that groups in Ψ_1 have cohomological dimension ≤ 2 and that groups in Ψ_2 have cohomological dimension ≤ 3. The above observations give no information about the word problem for groups of cohomological dimension 2. Collins and Miller (unpublished) have verified that some of the groups constructed in the Boone-Britton proofs of the unsolvability of the word problem have cohomological dimension 2.

THEOREM 4.12. *There exists a finitely presented group G of cohomological dimension 2 having unsolvable word problem. Indeed, G can be obtained from a free group by applying three successive HNN-extensions where the associated subgroups are finitely generated free groups.*

Of course the associated subgroups in the second and third HNN-extension are only free subgroups of the previous stage in the construction, not subgroups of the original free group. The proof is obtained by making minor variations to the one given in Rotman's textbook [92]. There a group $G = G(T)$ is constructed based on a Turing machine T which is first encoded into a semi-group and that in turn into G. The group G is obviously obtained by successive HNN-extensions. The only difficulty is to check that the associated subgroups in the final HNN-extension can be taken to be free. This can be done by arranging for the Turing machine T and the semigroup constructed to have a few special properties. The proof then uses the technical details of the proof in [92] and appeals to the deterministic nature and special properties of T.

5. Decision problems in algebraic classes of groups

While the fundamental decision problems are unsolvable for finitely presented groups in general, it is interesting to ask whether they can be solved for classes of finitely presented groups enjoying a particular algebraic property.

For example, it is easy to see that for the class of finite groups the word problem, the conjugacy problem, the generalized word problem and the isomorphism problem are all recursively solvable. Given any finite presentation of a finite group G, by enumerating Tietze transformations of the given presentation one can effectively find a multiplication table presentation for G. Each of the fundamental decision problems can then be effectively answered from knowledge of the multiplication table. Of course these are not practical algorithms and there is considerable interest in obtaining efficient practical algorithms for studying finite groups.

Similarly for the class of finitely generated free groups, the word and conjugacy problems are solvable by standard facts about equality and conjugacy. That the generalized word problem is solvable is a theorem of Nielsen [71]. The isomorphism type of a free group is determined by its rank which can be easily computed from any presentation by considering its abelianization. Hence all of the basic decision problems are solvable for finitely generated free groups.

In this section we survey what is known about the fundamental decision problems for classes of groups enjoying some of the more familiar algebraic properties, for example abelian, solvable, linear and so on. A diagram is included which summarizes the status of four fundamental problems and indicates some of the relationships between the various classes. Generally a class C_1 of groups is connected by a line to a class C_2 higher in the diagram if $C_1 \subseteq C_2$. Unfortunately not all containments and intersections can be accurately portrayed in the diagram.

The following notation is used for the various decision problems in the diagram: $+WP$ means that all groups in the class have solvable word problem; $\neg WP$ means that there exist groups in the class having unsolvable word problem; and $?WP$ means the solvability of the word problem seems to be an open question. Of course if a decision problem is solvable for a class of groups then it is solvable in every class contained that class. Likewise if a decision problem is unsolvable for a class of groups then it is also unsolvable for every larger class.

The following commentary, references and quoted results are intended to explain the status of the decision problems as indicated in the table.

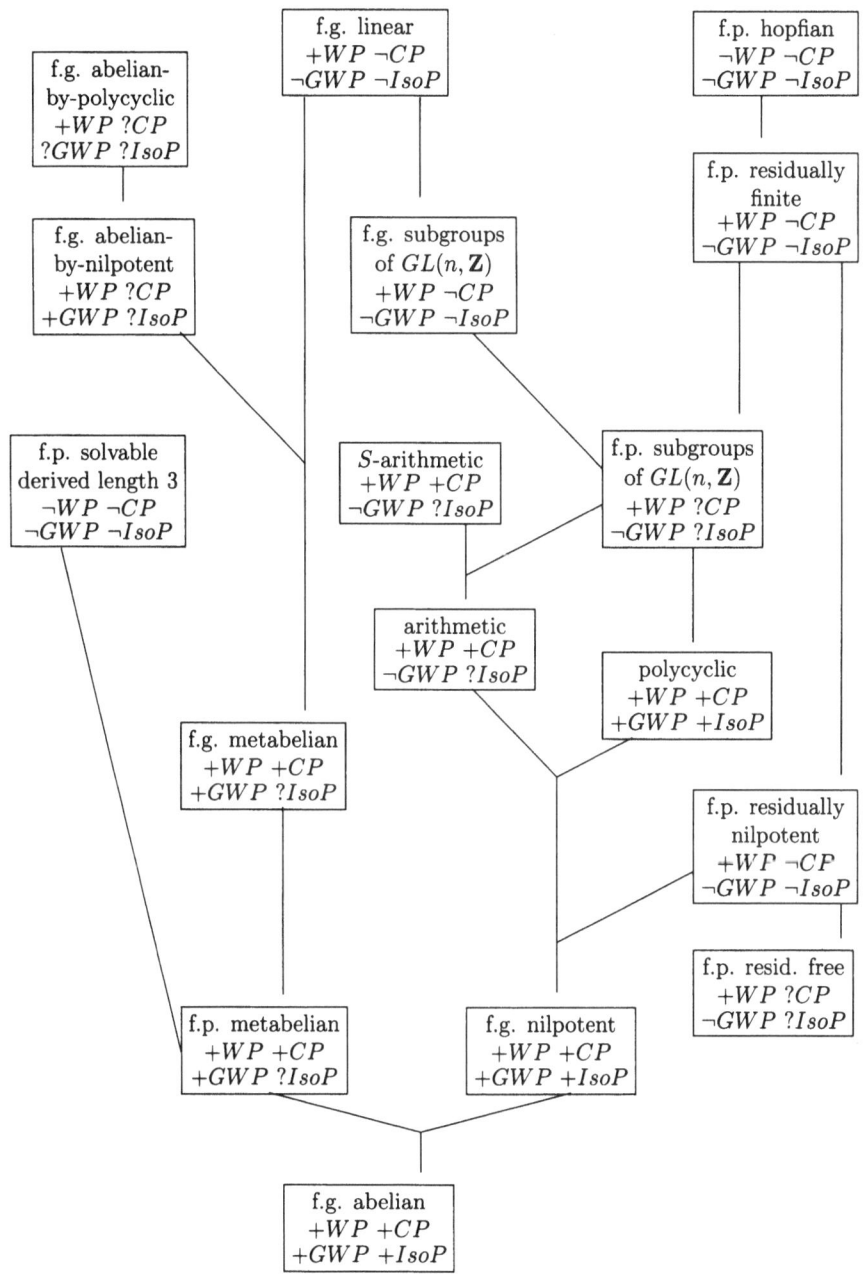

As we have mentioned before, the structure of a finitely generated abelian group can be completely and effectively determined from a finite presentation for such a group. In particular this enables one to solve the word problem in each such group and to solve the ismorphism problem for the class of such groups. Now for abelian groups conjugacy is the same as equality so the conjugacy problem is also solvable. The generalized word problem for a subgroup H of an abelian group G is equivalent to the word problem for G/H so $GWP(G)$ is also solvable. Thus for finitely generated abelian groups all of the basic decision problems are solvable.

THEOREM 5.1. *Finitely generated linear groups have solvable word problem.*

For finitely generated groups which are linear over a field, this result is proved in Rabin [89] but also follows easily from an older result of Malcev [72]. For Malcev proved that if a finitely generated group G has a faithful representation as a group of matrices over a field, then G also has a faithful representation over a field which is a purely transcendental extension of finite degree of the prime field. As the arithmetic of such a field is clearly effective, the word problem for such a group G is solvable.

More generally, a finitely generated group G of matrices with entries from a commutative ring has a solvable word problem. For since G is finitely generated, the entries in its matrices all lie in some finitely generated commutative ring. Now the arithmetic of such a ring is effective and hence such a group has solvable word problem (see [11]).

In the group $SL(2, \mathbf{Z})$ the two matrices

$$\begin{pmatrix} 1 & 2 \\ 0 & 1 \end{pmatrix} \quad \text{and} \quad \begin{pmatrix} 1 & 0 \\ 2 & 1 \end{pmatrix}$$

freely generate a free subgroup of rank 2. Moreover, Sanov [93] has shown that an arbirtary 2×2 matrix

$$\begin{pmatrix} a & b \\ c & d \end{pmatrix}$$

with integer entries belongs to this subgroup if and only if the following three arithmetic conditions are satisfied:
(1) $ad - bc = 1$
(2) a and d are congruent to 1 mod 4
(3) c and b are even.

It follows that the direct product $F_n \times F_n$ of two free groups of rank n has a faithful representation in $SL(4, \mathbf{Z})$. Combining this representation with the results discussed in the last section, one has the following:

THEOREM 5.2. *The conjugacy problem, the generalized word problem and the isomorphism problem are all unsolvable for finitely generated subgroups of $SL(4, \mathbf{Z})$. Further, the generalized word problem is unsolvable for arithmetic groups.*

Note that the finitely generated groups in this theorem need not be finitely presented. Such a group can however be finitely described just by giving the finite set of integer matrices which generate the group.

Recall that a group G is hopfian if $G/N \cong G$ implies $N = \{1\}$. Malcev [72] showed that finitely generated linear groups are residually finite. He further showed that finitely generated residually finite groups are hopfian.

Using the idea of "finite reducibility" which goes back to J. C. C. McKinsey [74] and was applied to groups by Dyson [33] and Mostowski [80] one can show the following result.

THEOREM 5.3. *Finitely presented, residually finite groups have solvable word problem.*

PROOF: (Sketch) For suppose we are given a finite presentation of a residually finite group G. To decide whether an arbitrary word w is equal to 1 in G we effectively enumerate two lists. The first list consists of all of the words equal to 1 in G, that is, all words which are freely equal to products of conjugates of the given defining relations.

The second list is more complicated: first systematicly enumerate all multiplication table presentations of finite groups. For any function f from the given generators of G to such a finite group K, one can effectively check whether f defines a homomorphism by checking to see whether the formal extension of f sends each of the finitely many defining relations of G to 1 in K. Thus we can effectively enumerate all homomorphisms f_i from G into finite groups K_i.

Now to decide whether $w =_G 1$ we start both listing processes. As each f_i is enumerated, evaluate $f_i(w)$ and check to see whether $f_i(w) = 1$ in K_i. Since G is residually finite, if $w \neq_G 1$ then for some i one eventually finds $f_i(w) \neq 1$ in K_i. On the other hand if $w =_G 1$ then w will appear in the first list. So by waiting until one of these two events occurs we can decide whether $w =_G 1$.

Despite the solvability of the word problem for finitely presented, residually finite groups the results explained in the previous section show the other fundamental decision problems are all unsolvable.

THEOREM 5.4. *The conjugacy problem, the generalized word problem and the isomorphism problem are all unsolvable for finitely presented, residually finite groups.*

Recently, Baumslag [9] has varied these constructions to show that these problems are also unsolvable for the class of finitely presented, residually nilpotent groups.

In view of Malcev [72] one can ask whether more generally finitely presented hopfian groups have solvable word problem. That hopfian groups can have unsolvable word problem follows easily from an embedding theorem of Miller and Schupp [79].

THEOREM 5.5. *A finitely presented group can be embedded into a finitely presented, hopfian group. In particular, there exist finitely presented hopfian groups with unsolvable word problem.*

One might have hoped that finitely presented solvable groups would have reasonable algorithmic properties. Any such hopes were destroyed by Kharlampovich [58] who constructed a finitely presented solvable group of derived length 3 with unsolvable word problem. Analysing and varying her construction, Baumslag, Gildenhuys and Strebel [18] have shown that the isomorphism problem is unsolvable for such groups. In summary:

THEOREM 5.6. *The word problem and the isomorphism problem are unsolvable for finitely presented solvable groups of derived length 3.*

In contrast to this situation for solvable groups in general, a large number of decision problems are recursively solvable for polycyclic groups. A general reference for polycyclic groups is the book by Segal [98]. First it is clear that polycyclic groups are finitely presented and have solvable word problem since these are poly–properties. The conjugacy problem is solvable for polycyclic groups since Remmeslenikov [91] and Formanek [34] have shown they are conjugacy separable, that is, two elements of a polycyclic group which are non–conjugate remain non–conjugate in some finite quotient group. Malcev [73] has shown that polycyclic groups are subgroup separable which implies the generalized word problem is solvable. Alternatively, the generalized word problem for polycyclic groups can be solved by a direct inductive method (see [12]).

Quite remarkably, Grunewald and Segal [48] have shown that the isomorphism problem for finitely generated nilpotent groups is recursively solvable. And more recently Segal [99] has succeeded in solving the isomorphism problem for polycyclic groups. In fact all of the algorithms mentioned carry over to the larger class of polycyclic–by–finite groups. In summary:

THEOREM 5.7. *The word problem, conjugacy problem, generalized word problem and isomoprphism problem for polycyclic–by–finite groups are recursively solvable.*

In fact a large number of other algorithmic questions about polycyclic–by–finite groups admit a positive solution (see [15]). Further, there is an algorithm to determine whether or not a given finitely presented solvable group is polycyclic (see [14]).

As part of their work leading to the solution of the isomorphism problem for finitely generated nilpotent groups, Grunewald and Segal solved the conjugacy problem for arithmetic groups (see [48]). Subsequently they extended this to S-arithmetic groups ([49]). They are further able to give a number of algorithms for determining orbits and constructing systems of generators for such groups.

THEOREM 5.8. *The conjugacy problem for S-arithmetic groups is recursively solvable.*

Next we turn our attention to finitely generated, abelian–by–polycyclic groups. In [51] Hall has shown that if G is a polycyclic group, then any finitely generated right $\mathbf{Z}G$-module is Noetherian. Of course this is equivalent to the assertion that $\mathbf{Z}G$ is right Noetherian and is an extension of the Hilbert basis theorem. Now a finitely generated, abelian–by–polycyclic group E is an extension of normal abelian subgroup M by a polycyclic group G. Then M can be viewed as a $\mathbf{Z}G$-module which must be finitely generated as a module because E is a finitely generated group. A particular instance of this is the case of finitely generated metabelian groups E where $M = [E, E]$ and $G = E/[E, E]$. When E is metabelian $\mathbf{Z}G$ is actually a finitely generated commutative ring.

In studying such groups one is able to apply the techniques of commutative algebra including the Hilbert basis theorem, the Nullstellensatz and so on. The advantage for algorithmic questions is that large parts of commutative algebra can be carried out effectively. In particular, there is an effective version of the Hilbert basis theorem which carries over to group rings $\mathbf{Z}G$ where G is a polycyclic group. Such group rings are *submodule computable* which roughly means that operations in their finitely generated modules can be effectively carried out, that one can decide membership in finitely generated submodules and that one can find presentations for finitely generated submodules. See [11] for an account of these effective methods.

One easy consequence of these effective extensions of commutative algebra is the solvability of the word problem for finitely generated, abelian–by–polycyclic groups. The generalized word problem is apparently more difficult, but using Theorem 2.14 of [11] one can solve the generalized word problem for finitely generated, abelian–by–nilpotent groups. The problem is reduced to deciding whether two submodules over the same subgroup of a nilpotent group coincide. Whether this can be extended to the abelian–by–polycyclic case seems to be unknown. In summary:

THEOREM 5.9. *The word problem for finitely generated, abelian–by–poly–cyclic groups is recursively solvable. The generalized word problem for finitely generated, abelian–by–nilpotent groups is recursively solvable.*

The particular case of metabelian groups is even more tractable. For finitely generated metabelian groups, Noskov [85] has solved the conjugacy problem. In summary:

THEOREM 5.10. *For finitely generated metabelian groups the word problem, generalized word problem and conjugacy problem are all recursively solvable.*

Finitely generated metabelian groups need not be finitely presented. But from an algorithmic point of view there seems to be little advantage in assuming finite presentation. Decision problems are reduced to questions about modules over finitely generated commutative rings. The isomorphism problem for finitely generated metabelian groups seems to be open. Indeed the following important and related algorthmic problem seems to be open: Is the isomorphism problem for finitely presented modules over the integral polynomial ring $\mathbf{Z}[x_1, \ldots, x_n]$ recursively solvable?

This completes our commentary concerning the summary table. Decision problems for certain other algebraic classes will be considered in a subsequent section.

6. Algorithms for further classes of groups

In this section we review the status of the fundamental decision problems for some further classes of groups which did not fit conveniently into the previous two sections. Nevertheless algorithmic questions concerning these groups have played an important role in ongoing developments.

One relator groups: Consider a group G defined by a single defining relation, say

$$G = < x_1, \ldots, x_n \mid r = 1 >$$

where r is a cyclically reduced word on the x_i. Magnus [69] (see also [71] or [66]) initiated the study of such one relator groups by proving his Freiheitsatz: the subgroup of G generated by a subset $S = \{x_{i_1}, \ldots, x_{i_k}\}$ of the given generators which omits at least one generator appearing in the cycllically reduced word r is a free group with basis S. His proof was by induction on the length of the defining relation r using a rewriting method which has become a powerful technique for studying one relator groups. Using this technique Magnus [70] succeeded in solving the word problem for such groups in the following strong sense: there is an algorithm to determine for an arbitrary word w of G and an arbitrary subset $S = \{x_{i_1}, \ldots, x_{i_k}\}$ of the given generators whether or not w belongs to the subgroup generated by S. Of course the ordinary word problem is just the case $S = \emptyset$. Note however that one can only decide membership in subgroups of a certain form. The full generalized word problem for one relator groups seems to be open.

In the case the defining relation is a proper power, say $r = u^n$ for some $n > 1$, Newman [84] proved a "Spelling Theorem" which provides a very sharp solution to the word problem (see [66]). Using this result he further solved the conjugacy problem for these one relator groups with torsion. Quite recently, Juhasz [57] has succeeded in solving the conjugacy problem for all one relator groups. In summary:

THEOREM 6.1. *The word and conjugacy problems are recursively solvable for groups defined by a single defining relation.*

The isomorphism problem for one relator groups is open. Some modest progress has been made on classifying one relator groups with relation of a particular form, but progress seems difficult.

Simple groups: In [63] Kuznetsov observed the following result which holds more generally for a large number of algebraic systems.

THEOREM 6.2. *A recursively presented simple group G has solvable word problem.*

PROOF: Suppose $G = < x_1, \ldots, x_n \mid r_1 = 1, r_2 = 1, \ldots >$. If $G = 1$ the result certainly holds. So assume $G \neq 1$ and let u be a fixed word of G such that $u \neq_G 1$. Now for any word w on the generators of G, let G_w be the group obtained from G by adding w as a new defining relator, that is,

$$G_w < x_1, \ldots, x_n \mid w = 1, r_1 = 1, r_2 = 1, \ldots > .$$

Now if $w \neq_G 1$ then $G_w = 1$ since G is simple and in particular $u = 1$ in G_w. But if $w =_G 1$ then $G_w = G$ and of course $u \neq 1$ in G_w. Clearly G_w is again recursively presented.

To decide whether an arbitrary word w is equal to 1 in G begin recursively enumerating two lists of words. The first list consists of all word equal to 1 in G. The second list consists of all words equal to 1 in G_w. If $w =_G 1$ then w will appear in the first list. But $w \neq_G 1$ if and only if u appears in the second list. By examining the lists until one of these events occurs we can determine whether or not w is equal to 1 in G. This completes the proof.

In particular, finitely presented simple groups have solvable word problem. In contrast to this, Scott [94] has shown the following:

THEOREM 6.3. *There is a finitely presented simple group whose conjugacy problem is recursively unsolvable.*

See [95] for a recent survey concerning finitely presented infinite simple groups. Boone and Higman [23] have used a variation on Kuznetsov's argument and the Higman embedding theorem to give the following characterization of finitely presented groups with solvable word problem.

THEOREM 6.4 (Boone-Higman Theorem). *A finitely presented group G has solvable word problem if and only if G can be embedded in a simple subgroup of a finitely presented group.*

PROOF: (Sketch) Suppose $G \subseteq S \subseteq H$ where S is simple and H is finitely presented. Fix a word $u \in S$ with $u \neq_H 1$. For any word w of H let H_w be the presentation obtained from that of H by adding w as a new defining relator. Now to decide whether a word in the generators of G is equal to 1 in G we regard it as a word, say w, in the generators of H. As above, either w appears on the list of words equal to 1 in H or else since S is simple u appears on the list of words equal to 1 in H_w (in which case $w \neq_H 1$). So by enumerating these two lists we can decide whether w is equal to 1 in G.

For the converse, suppose G has solvable word problem. Then the set of all pairs of words (u, v) such that $u \neq_G 1$ and $v \neq_G 1$ is recursive and can be arranged in a recursive list as say $(u_i, v_i), i = 1, 2, \ldots$. Let $x, t_1, t_2, \ldots$ be new generating symbols and form the presentation

$$\sigma(G) = < G, x, t_1, t_2, \ldots \mid t_i^{-1}[u_i, x]t_i = v_i x^{-1} u_i x, i = 1, 2, \ldots > .$$

Then $\sigma(G)$ is an HNN-extension of the free product of G with the infinite cyclic group generated by x. The associated subgroups are just the infinite cyclic groups genrated by the $[u_i, x]$ and the $v_i x^{-1} u_i x$. A routine argument shows that the word problem for $\sigma(G)$ in the indicated presentation can be solved using the given solution to the word problem for G. Also observe that in $\sigma(G)$ each v_i lies in the normal closure of the corresponding u_i. Thus the normal closure in $\sigma(G)$ of any non-trivial element of G contains all of G.

Now iterating this construction $G_1 = \sigma(G), G_2 = \sigma(G_1), \ldots$ and forming the union

$$S = \bigcup_{j=1}^{\infty} G_j$$

we obtain a recursively presented simple group S in which G is embedded. By the Higman embedding theorem, S can be embedded in a finitely presented group H. This completes the proof.

Small cancellation groups: A subset R of a free group F is *symmetrized* if all of the elements of R are cyclically reduced and if $r \in R$ implies that all cyclically reduced conjugates of $r^{\pm 1}$ are also in R. Thus the words in R are all cyclically reduced and are closed under taking inverses and cyclic permutations. Let $N =< R >^F$ be the normal closure of R in F. Clearly any presentation (respectively, finite presentation) of a group can be converted to a presentation on the same set of generators with a symmetrized set (respectively, finite symmetrized set) of defining relators. One just cyclically reduces the given relators and then closes under inverses and cyclically permutations of words.

In *small cancellation theory* one consider various cancellation hypotheses on a symmetrized set of words R and uses them to deduce properties of the group $G = F/N$. In what follows we assume R is a symmetrized set of words.

If R contains two distinct words of the form $r_1 \equiv bc_1$ and $r_2 \equiv bc_2$ then the word b is called a *piece relative to R* or simply a *piece* when R is understood. Observe that, in forming the product $r_1^{-1} r_2$ and freely reducing, such a piece b is cancelled. Thus a piece is simply a subword of an element of R which can be cancelled by the multiplication of two non-inverse elements of R. Also note that an initial segment of a piece is again a piece.

There are two types of small cancellation hypotheses which assert that pieces are relatively small parts of elements of R. The first is a *metric condition* denoted $C'(\lambda)$ where λ is any positive real (for example $\frac{1}{6}$ or $\frac{1}{8}$). The set R is said to satisfy $C'(\lambda)$ if $r = bc \in R$ where b is a piece implies that $|b| < \lambda|r|$. For instance, $C'(\frac{1}{6})$ means that in forming the product of any two non-inverse elements of R less that $\frac{1}{6}$ of either word is cancelled. Note that $C'(\frac{1}{8})$ implies $C'(\frac{1}{6})$ and that generally $C'(\lambda)$ is a stronger condition for smaller λ.

For any natural number p the *non-metric condition* $C(p)$ asserts that no element of R is a product of fewer than p pieces. Observe that $C'(\lambda)$ implies $C(p)$ for $\lambda \leq 1/(p-1)$. Thus $C'(\frac{1}{6})$ implies $C(7)$.

Another type of condition considered in small cancellation theory is the condition $T(q)$ for q a natural number. R satisfies $T(q)$ if for every sequence $r_1, \ldots, r_m$ ($3 \leq m \leq q$) with no successive inverse pairs, at least one of the products $r_1 r_2, \ldots, r_{m-1} r_m, r_m r_1$ is reduced without cancellation. (This condition turns out to be dual to to the condition $C(p)$ when $1/p + 1/q = 1/2$ in a suitable geometric sense. See [66].)

Small cancellation theory was initiated by Tartakovskii [101] [102] [103] who solved the word problem for groups whose defining relators R satisfy $C(7)$. Greendlinger [43] investigated the metric conditions $C'(\lambda)$ and showed that the word problem for groups with defining relators R satisfying $C'(\frac{1}{6})$ is solvable by *Dehn's algorithm* (see below). In addition he obtained quite a strong result for such groups called "Greedlinger's Lemma" which has a number of applications (see [66]). Further, Greendlinger [44] showed that the conjugacy problem for groups with defining relators R satisfying $C'(\frac{1}{8})$ is solvable by *Dehn's conjugacy algorithm* (see below).

Lyndon [65] introduced the geometric method of *Lyndon-van Kampen diagrams* into small cancellation theory and solved the word problem for groups whose defining relators satisfy $C(6)$ (and also in certain other cases). Schupp [96] used these geometric methods to obtain a solution to the conjugacy problem for these groups. In summary:

THEOREM 6.5. *Let F be a free group, R a finite symmetrized subset of F and N the normal closure of R. Assume that R satisfies either $C(6)$, or $C(4)$ and $T(4)$, or $C(3)$ and $T(6)$. Then the word problem and the conjugacy problem for $G = F/N$ are recursively solvable.*

There are many applications of small cancellation theory and much more detailed information than we can present here. Moreover, the geometric methods using Lyndon-van Kampen diagrams provide significant insight into the word and conjugacy problems. A very readable account of small cancellation theory is given in Chapter V of the book [66] by Lyndon and Schupp.

Varieties of groups: In the previous section we noted Kharlampovich's result [58] that there is a finitely presented solvable group of derived length 3 with unsolvable word problem. In particular, this is an example of a finitely presented group satisfying a non-trivial varietal law having unsolvable word problem.

For groups in a non-trivial variety one often considers groups that are *relatively finitely presented*, that is, finitely generated groups which are defined by the laws of the variety together with finitely many additional relations. For example, a result of Hall [51] shows that finitely generated metabelian groups are always relatively finitely presented although they need not be "absolutely" finitely presented.

In a series of papers Kleiman [59], [61], [60], [62] has proved a number of remarkable results concerning varieties which answer a large number of questions. We mention only some of those results of an algorithmic nature.

THEOREM 6.6 ([59], [61]). *There is a solvable variety defined by finitely many laws in which the non-cyclic free groups have unsolvable word problem.*

An easy consequence of the unsolvability of the word problem for these relatively free groups is the following:

COROLLARY 6.7. *There is a finite set S of identities such that the problem to decide whether an arbitrary word represents an identity which is a consequence of S is recursively unsolvable.*

As another consequence, Kleiman shows that the problem of recognizing whether a variety can be factored into a product is recursively unsolvable. In the context of varieties, the *identity problem* for a group is the problem of deciding whether or not an arbitrary identical relation (law) holds in the group. Kleiman [61] has also shown that there is a 3 generator group with solvable word problem in which the identity problem is unsolvable.

7. Geometry and complexity

Suppose that the group G has the presentation $P = < X \mid R >$ where R is a symmetrized set of words on X. Let F be the free group with basis X and N the normal closure of R. If w is a word in the group $G = F/N$ then $w =_G 1$ if and only if $w =_F u_1 r_1 u_1^{-1} \cdots u_m r_m u_m^{-1}$ where the u_i are words in F and the r_i are elements of R. The sequence $u_1 r_1 u_1^{-1}, \ldots, u_m r_m u_m^{-1}$ is said to be an R-sequence of length m for w. For any $w \in N$ we define $A_P(w)$ to be the minimium length of an R-sequence for w.

Using Lyndon-van Kampen diagrams one associates with any R-sequence for w a connected, simply connected R-*diagram* D (consisting of vertices, edges and regions) in the euclidean plane. The edges of D are labelled by elements of F in such a way that the label on the boundary of each region is an element of R and the label on the boundary circuit of D is w (see Chapter V of [66]). Conversely, R-diagrams with boundary label w are the diagrams of suitable R-sequences for w. Thus the $A_P(w)$ is the number of regions in the R-diagram of a minimal R-sequence for w.

For any word u of F we denote the length of u by $|u|$. A useful observation that is easily established using R-diagrams is the following (see [66] Lemma 1.2 of Chapter V, p. 239):

LEMMA 7.1. *If $w \in N$ and $A_P(w) = m$ then there is an R-sequence*

$$u_1 r_1 u_1^{-1}, \ldots, u_m r_m u_m^{-1}$$

of length m for w such that

$$|u_j| \leq |w| + \sum_{i=1}^{m} |r_i| \quad j = 1, \ldots, m.$$

In particular, if the lengths of all the elements of R are bounded by some constant, say C_R, then the conjugating elements u_j in a minimal R-sequence can be chosen so that $|u_j| \leq |w| + m \cdot C_R$.

Consequently, when R is finite, in order to decide whether a word w lies in N or not, it suffices to have an upper bound for $A_P(w)$ in terms of $|w|$. For then one can systematically try the bounded number of R-sequences with conjugating elements of length at most the above estimate. Indeed it is easy to see that being able to compute such a bound is equivalent to solving the word problem in the case that R is finite.

COROLLARY 7.2. *Suppose that G has finite presentation $P = < X \mid R >$ where R is a symmetrized set of words on X. Let F be the free group with basis X and N the normal closure of R. Then the word problem for $G = F/N$ is recursively solvable if and only if there is a recursive function f such that $A_P(w) \leq f(|w|)$ for all $w \in N$.*

It is often helpful to use a more invariant form of the function A_P. Thus one defines [36] the *Dehn function* Ω_P as follows:

$$\Omega_P(n) = max\{A_P(w) \mid w \in N \text{ and } |w| \leq n\}.$$

It can be shown that if the finitely presented group G is defined by two finite presentations P and P_1, then there are constants c_1, c_2 and c_3 such that $\Omega_{P_1}(n) \leq c_1 \Omega_P(c_2 n) + c_3 n$. In a different direction, suppose that the finite presentation $P_2 = < X \mid R_2 >$ of G is obtained from P by adding some elements of N to R so that $R \subseteq R_2$. Then $A_{P_2}(w) \leq A_P(w)$ for all $w \in N$ and $\Omega_{P_2}(n) \leq \Omega_P(n)$.

The above corollary can be restated now as follows: the word problem for G is solvable if and only if there is a recursive function f such that

$\Omega_P(n) \leq f(n)$ for all $n > 0$. In fact, if Ω_P is bounded by such a recursive function, it then follows that Ω_P itself is recursive. So the corollary becomes: the word problem for G is solvable if and only if Ω_P is recursive.

Assuming the presentation $P = < X \mid R >$ is finite, there is a standard and familiar way to realize $G = F/N$ as the fundamental group of a 2-complex $K = K(P)$ consisting of a single 0-cell, one 1-cell for each free generator of F, and one 2-cell for each element of R. The 1-cells are attached as loops at the 0-cell giving a wedge of circles. The 2-cells are attached to the 1-skeleton by subdividing the boundary of the cell corresponding to $r \in R$ and sewing onto the 1-skeleton in accordance with r as a sequnce of generators and their inverses. The Seifert-van Kampen Theorem then shows that $\pi_1(K) \cong G$.

If an element $w \in N$ is represented as a loop $\tau(w)$ in the 1-skeleton of K and if D is the R-diagram of a (minimal) R-sequence for w, then there is clearly a continuous map from D to K which sends the boundary of D to $\tau(w)$, sends the edges of D into the 1-skeleton of K and sends the regions of D onto 2-cells of K. Since D is connected and simply connected this map lifts to a map of D into the universal covering space $\widetilde{K}$.

If we fix a 0-cell in the universal cover $\widetilde{K}$ as base point, then the 0-cells of $\widetilde{K}$ are in one-one correspondence with the elements of G. Now the 1-skeleton of $\widetilde{K}$ is a graph which we denote by $\Gamma(G)$ called the *Cayley graph of* G. Another way to view $\Gamma(G)$ is as follows: regard F as the fundamental group of the 1-skeleton K^1 of K. Then $\Gamma(G)$ can be identified with the covering space of K^1 corresponding to the normal subgroup N of F. Note that $\Gamma(G)$ depends only on the choice of generating set X and not on the choice of the defining relators. We write $\Gamma(G) = \Gamma(G, X)$ to show this dependence when necessary. Thus we think of the vertices of $\Gamma(G)$ as being labelled by the elements of G. For each generator in X there is an oriented edge entering and an edge leaving each vertex of $\Gamma(G)$. Paths in $\Gamma(G)$ correspond to words in X starting at the vertex labelled 1. In particular, loops starting at the vertex labelled 1 correspond to elements of N.

The Cayley graph $\Gamma(G) = \Gamma(G, X)$ is given the *word metric* defined by taking each edge to have unit length. A *geodesic* for an element $g \in G$ is a shortest path w in the Cayley graph from the vertex labelled 1 to the vertex labelled g. This w is a shortest word in the generators X representing the element G.

Dehn's algorithm and hyperbolic groups: Let G be a group with finite presentation $P = < X \mid R >$ and let F and N be as above. A *Dehn's algorithm for* G is a finite set of words $\Delta \subset N$ such that any non-empty word $w \in N$ can be shortened by applying a relator in Δ. That is, any non-empty $w \in N$ has the form $w \equiv ubv$ where there is an element of the form $bc \in \Delta$ with $|c| < |b|$. Thus applying the relator bc to w, we can deduce that $w =_G uc^{-1}v$ where $|uc^{-1}v| < |w|$.

If G has Dehn's algorithm Δ then one can solve the word problem for G in a particularly straight forward way: repeatedly try to shorten the word in question by replacing a subword using a relator in Δ. If Δ is a Dehn's algorithm for G as above, then $< X \mid \Delta >$ is also a finite presentation for G and $A_{<X \mid \Delta>}(w) \le |w|$ for every $w \in N$. Thus also $\Omega_{<X \mid \Delta>}(n) \le n$ for all n, so the Dehn function is bounded by a linear function. Observe this last property does not depend on the finite presentation of G.

If Δ is a Dehn's algorithm for G and Δ_1 is a larger finite set of words with $\Delta \subseteq \Delta_1 \subset N$, then Δ_1 is also a Dehn's algorithm for G. In particular, if c is a constant larger than the lengths of all the words in Δ then

$$\Delta_c = \{w \in N \mid |w| \le c\}$$

is also a Dehn's algorithm. Using these observations and Theorem 7.3 below, one can check that having a Dehn's algorithm is independent of the choice of generating set and hence is an abstract property of the group.

The following are examples of groups having presentations with a Dehn's algorithm: free groups, finite groups (multiplication table presentation), and groups satisfying the cancellation condition $C'(\frac{1}{6})$ (Greendlinger's Lemma).

After introducing the fundamental decision problems, Dehn [31] considered the fundamental groups of closed orientable surfaces of genus $g > 1$ having presentation

$$S_g =< a_1, b_1, \ldots, a_g, b_g \mid [a_1, b_1] \cdots [a_g, b_g] = 1 > .$$

He showed that the symmetrized closure R of the given defining relator is a Dehn's algorithm for S_g. Note that this presentation satisfies the small cancellation coditions $C(4g)$ and $C'(1/(4g - 1))$.

Suppose the group G has a finite presentation $< X \mid R >$ where R is a Dehn's algorithm. A word u is said to be R-reduced if u can not be shortened by applying a relator in R. Also, u is cyclically R-reduced if every cyclic permutation of u is R-reduced. Clearly if u is any word on X, then by taking cyclic permutations and applying Dehn's algorithm one can effectively find a cyclically R-reduced word u' which is conjugate to u in G.

We say that Dehn's conjugacy algorithm solves the conjugacy problem for G if there is an integer constant c such that two non-trivial cyclically R-reduced words u and v are conjugate in G if and only if there is a word z on X such that $u =_G z^{-1}vz$ and $|z| \le c \cdot (|u| + |v|)$. This clearly provides a recursive solution to the conjugacy problem for G since only finitely many conjugating elements z need to be tested. Also note that the number of regions in an annular R-diagram representing such a conjugacy equation is bounded by a linear function of $(|u| + |v|)$.

Dehn also solved the conjugacy problem for the groups S_g by showing what we have called Dehn's conjugacy algorithm applies. Actually, for S_g and more generally $C'(\frac{1}{8})$ groups a stronger condition holds. Namely, if u and v as before are conjugate, then they have cyclic permutations u' and v' such that $u' =_G z^{-1}v'z$ where z is a subword of an element of R. See [66] for details and generalizations to infinite sets of defining relators.

Gromov [45] has given several equivalent definitions of the notion of a *word hyperbolic group*. Let G be a finitely generated group and fix a finite generating set X for G. Let $\Gamma = \Gamma(G, X)$ be the corresponding Cayley graph with the word metric.

A triangle in Γ with geodesic sides is said to be δ-*thin* if any point on one side is at distance less than δ from some point on one of the other two sides. Γ is said to be δ-*hyperbolic* if every triangle in Γ with geodesic sides is δ-thin. Finally G is said to be *word hyperbolic* if it is δ-hyperbolic with respect to some generating set and some fixed $\delta \geq 0$. We remark that one should think of δ as a large integer rather than a small positive real in these definitions.

The following result gives some equivalent characterizations of word hyperbolic groups.

THEOREM 7.3. *The following conditions on a finitely presented group G are equivalent:*
  (1) G is word hyperbolic (in the sense of δ-thin triangles);
  (2) the Dehn function for G is bounded by a linear function;
  (3) G has a Dehn's algorithm.

The above is a combination of results of Gromov [45] and of (independently) Lysënok [67] and Shapiro [5] (see also [8]). Not only do word hyperbolic groups have solvable word problem (by this theorem), but Gromov [45] has also shown they have solvable conjugacy problem:

THEOREM 7.4. *Let G be a finitely presented word hyperbolic group. Then Dehn's conjugacy algorithm solves the conjugacy problem for G.*

Thus Dehn's algorithm for the word problem always implies that Dehn's conjugacy algorithm solves the conjugacy problem. In terms of R-diagrams this means the following: if the number of regions in a R-diagram which is a disk is bounded by a linear function of the length of the boundary, then the same is true of annular R-diagrams.

Automatic groups: The notion of an *automatic group* was introduced in [26]. Another reference is [17].

We will need the notion of a *finite state automaton*. Intuitively, a finite state automaton M is just a computing device with a fixed finite amount of storage (memory). M reads an input string (or word) from a free monoid Φ on a finite alphabet from a tape and eventually either "accepts" or "rejects" the input string. M reads only in one direction (no backups) and can write only in its fixed internal storage. (If arbitrarily large storage for scratch work were allowed, the resulting class of machines would be as powerful as Turing machines)

The set of strings (words) which M accepts is called the language recognized by M. A *regular language* is a set of words in a free monoid Φ which is recognized by some finite state automaton.

To compare two words using a finite state automaton one pads the shorter with a new symbol, say $, on the end so the two words have the same length. Then intersperse these two words on the input tape taking symbols alternately from the two words. Equivalently, one can use a two tape automaton which reads its tapes with one (possibly padded) word on each at the same rate (synchronously). We call such a device a *synchronous two-tape automaton*. If instead such a two tape automaton is allowed to read its input tapes at different rates, we call such a device an *asynchronous two-tape automaton*. An asynchronous two tape automaton can recognize far more complicated sets of pairs than a synchronous one.

Let X be a set of generators for a group G. Let Φ be the free monoid with basis $X \cup X^{-1}$. For any word $w \in \Phi$ define $\mu(w) \in G$ to be the element represented by w. An *(synchronously) automatic structure for G* with respect to the generating set X is a regular language L in Φ such that $\mu(L) = G$ together with a synchronous two-tape automaton M which accepts the collection of pairs of elements of L which represent elements of G lying at most a unit apart in the Cayley graph $\Gamma(G, X)$. That is M accepts the set of pairs

$$\{(u, v) \mid u, v \in L \text{ and } \mu(u) = \mu(va) \text{ for some } a \in X \cup X^{-1} \cup \{1\}\}.$$

If in addition there is a finite state automaton M' which accepts the set of pairs

$$\{(u, v) \mid u, v \in L \text{ and } \mu(u) = \mu(bv) \text{ for some } b \in X \cup X^{-1} \cup \{1\}\}$$

the triple (L, M, M') is called a *biautomatic* or *two-sided automatic structure*. Finally, if instead M is an asychronously automaton, we say that (L, M) is an *asynchronously automatic structure*.

Note that while μ maps L onto G it need not be one-one; that is, and element of G may be represented by several elements of L.

The group G is said to be *automatic (respectively, biautomatic or asynchronously automatic)* if it has an automatic (respectively, biautomatic or asynchronously automatic) structure. Of course biautomatic groups are automatic, but it is not known whether these two notions coincide. Automatic groups are asynchronously automatic, but the class of asynchronously automatic groups is much larger.

One of the motivations for studying automatic groups was the observation [26] that word hyperbolic groups are automatic (even biautomatic [38]). Note that if R is a Dehn's algorithm for a word hyperbolic group, then the R-reduced words form a regular language. Moreover, it can be shown (see Gromov [45]) that if G is word hyperbolic with Cayley graph Γ, then the set of words on X corresponding to geodesic paths starting at $1 \in \Gamma$ is a regular language.

While not all of the non-metric small cancellation groups are word hyperbolic, Gersten and Short [37] have shown that groups satisfying any one of $C(6)$, or $C(4)$ and $T(4)$, or $C(3)$ and $T(6)$ have an automatic structure.

The word problem for these classes of automatic groups is solvable, and more detailed information is as follows (see [26] and [17]):

THEOREM 7.5. *Asynchronously automatic groups all have solvable word problem. The Dehn function of an automatic group is bounded by a quadratic. The Dehn function Ω of an asynchronously automatic group is bounded by a simple exponential, that is, $\Omega(n) \leq c^n$ for some constant $c > 0$.*

Moreover, Gersten and Short [37] have shown the following result which generalizes previously mentioned results of Schupp and of Gromov:

THEOREM 7.6. *The conjugacy problem for biautomatic groups is recursively solvable.*

Their proof uses various closure properties of regular languages and reformulates the conjugacy problem as a question about whether two regular languages have a non-empty intersection. This latter problem is known to be solvable [56].

For asynchronously automatic groups the conjugacy problem is no longer solvable. Indeed, it is not hard to show [17] that the split extension of one finitely generated free group by another is asychronously automatic. Consequently, the examples of [77] discussed in the earlier section on "Decision problems and constructions" show the following:

THEOREM 7.7. *There exist asynchronously automatic groups with unsolvable conjugacy problem. The isomorphism problem for asynchronously automatic groups is recursively unsolvable.*

For more information on asynchronously automatic groups see [17]. These various classes of (bi)automatic groups have number of other interesting properties. Thurston has shown that automatic groups are of type FP_∞ (see [4]). Gersten and Short [39] have obtained useful information about subgroups of biautomatic and hyperbolic groups.

Normal forms and rewriting systems: Continuing with the above notations, one difficulty with Dehn's algorithm for solving the word problem is the following: if R is a Dehn's algorithm and we apply the process of R-reduction to a word $w \neq_G 1$ the resulting word, say $\rho(w)$ will be R-reduced but it is not unique. There may be other R-reduced words u with $u =_G \rho(w)$; for instance, one might have applied a completely different sequence of R-reductions.

One would like a solution to the word problem which transforms an arbitrary word into a sort of unique standard form, preferably by straight forward operations. Also it seems reasonable to expect that a subword of a word in standard form is also in standard form; otherwise one should continue trying to transform the subwords. Notice that the R-reduction process has all of these properties except uniqueness.

A set of words T in the free monoid Φ with basis $X \cup X^{-1}$ is a *set of normal forms* for G (with respect to the generating set X) if $\mu|_T$ is a bijection from T onto G. Thus every element of G is represented by a unique element of T. Equivalently, one can view T as a transversal in F of the normal subgroup N. If in addition T is closed under taking of subwords, then we call T a *hereditary set of normal forms*.

Observe that a set of words is closed under taking subwords if and only if it is closed under taking both initial and terminal segments of words. Thus T is a hereditary set of normal forms if and only if it is a two-sided Schreier transversal for N in F. One can well-order the words in Φ by ordering $X \cup X^{-1}$ and then ordering Φ by length and within the same length lexicographically. In [50] M. Hall shows the set T obtained by choosing the least element t_g representing each group element $g \in G$ is a two-sided Schreier transversal.

Thus every group G has a hereditary set of normal forms T with respect to a generating set X. Assuming X is finite, the set T constructed in this way is recursive if and only if G has solvable word problem.

So normal forms exist, but what of "transforming" words into normal form? One such notion has been investigated by computer scientists (see

[64] or [68]). Define a *rewrite rule* to be an ordered pair (u, v) of words of
Φ such that $u =_G v$. An *application* of the rewrite rule (u, v) consists of
replacing a subword of the form u in a word w by v to obtain a new word
w'. We write this as $w \rightarrow w'$. In more detail, an application of the rewrite
rule (u, v) looks like $w \equiv yuz \rightarrow w' \equiv yvz$. One often writes $u \rightarrow v$ for
the rewrite rule itself thereby emphasizing the ordered nature of the rule.
The words u and v are called the left and right hand sides of the rewrite
rule $u \rightarrow v$ respectively. Note that an application of a rewrite rule need not
reduce the length of a word, and it may indeed lengthen the word.

Let Λ be a set of rewrite rules. A word w is Λ-*reduced* or Λ-*irreducible*
if no subword of w is the left hand side of any rewrite rule in Λ. That is, w
is Λ-irreducible if it is impossible to apply any of the rewrite rules in Λ to
W.

A set of rewrite rules Λ is a *complete rewriting system* if it satisfies two
conditions: (1) the set T of Λ-irreducible words is a hereditary set of normal
forms; and (2) there are no infinite chains $w_1 \rightarrow w_2 \rightarrow \ldots$ of applications
of rewrite rules from Λ.

If Λ is a complete rewriting system, it is clear that starting with any
word w one can apply successively the rewrite rules of Λ to reach a unique
normal form for w. If w_1 and w_2 are two words such that $w_1 =_G w_2$, then
applying rewrite rules to each of w_1 and w_2 in any order will eventually
lead to the same word. A rewrite system with this property is said to be
confluent, a condition which could have been used in place of (1). Also,
if Λ is a complete rewriting system then the collection of equations $u = v$
where $(u, v) \in \Lambda$ give a presentation for G.

Here is one (non-effective) way to obtain such a complete rewriting
system. Choose a hereditary set T of normal forms as before. If w is a
word of Φ, denote by $\overline{w}$ the unique element of T which is equal in G to
w (the coset representative for wN in F). Take Λ to be the set of rewrite
rules $ta \rightarrow \overline{(ta)}$ for all $t \in T$ and all $a \in X \cup X^{-1}$. We observe that if G
has solvable word problem then this set of rewrite rules is recursive.

Of particular interest is the class of finitely generated groups which have
a finite complete rewriting system. If G has a finite complete rewriting
system Λ, then G is finitely presented and the set T of Λ-irreducible words
is a regular language. The word problem for such a group is easily solved
by repeatedly applying the finite set of rewrite rules until an irreducible
word is obtained.

Examples of such groups are free groups and finite groups. One can
also show that the class of groups with finite complete rewriting systems
is closed under ordinary free products and under extensions (see [47]). In

particular, this class includes all polycyclic–by–finite groups and groups which are extensions of one finitely generated free group by another. So from our previous results we can summarize the status of the fundamental decision problems as follows:

THEOREM 7.8. *The word problem for groups with a finite complete rewriting system is recursively solvable. There exist groups groups with a finite complete rewriting system having unsolvable conjugacy problem and unsolvable generalized word problem. Moreover the isomorphism problem for such groups is unsolvable.*

It should be emphasized again that rewrite rules are not required to be length reducing. If a group G has a finite, length reducing, complete rewriting system, then that system gives a Dehn's algorithm and, moreover, it is known (see [68]) that G must be virtually free.

One interesting feature of groups with a finite complete rewriting system is that they are of type FP_∞ (see [6], [25], [46] and [100]). Moreover, one can in principle effectively calculate free resolutions and carry out certain homological calculations for such groups. The nature of subgroups of such groups remains to be explored, as do a number of generalizations. For instance, instead of finite systems one can consider "regular" complete rewriting systems. Exactly how these might be related to the various types of automatic groups is not yet clear.

Trivial words as a language: Another way of approaching the word problem is to consider the collection $N(\Phi)$ of all words $w \in \Phi$ such that $w =_G 1$. Of course words in $N(\Phi)$ represent elements of N but they may not be freely reduced. One measure of the complexity of the word problem for G is the complexity of $N(\Phi)$ as a language. Thus a group G is said to be *regular* (respectively, *context-free*) if $N(\Phi)$ is a regular (respectively, context–free) language.

Recall that a language is *context-free* if it is recognized by a pushdown automaton which is a non-deterministic finite state automaton with a pushdown stack ("first in, last out") storage device. See [56] for details concerning such languages and machines.

The following result was observed by Anisimov [7]. Its proof is an easy exercise (see [81]).

THEOREM 7.9. *A group is regular if and only if it is finite.*

In a series of papers, Muller and Schupp [81], [82] investigate some remarkable connections between groups, pushdown automata, the theory of ends and second-order logic. One of their results [81] is a characterization of context–free groups which involves the notion af accessibility. Subsequently Dunwoody [32] has proved that all finitely presented groups are accessible. So combining these results, one has the following:

THEOREM 7.10. *A finitely generated group is context–free if and only if it is virtually free.*

For additional information about rewriting systems and complexity issues for groups the reader may wish to consult the survey article [68].

8. Computability of homological invariants

This section is concerned with decision theoretic aspects of the homological invariants of finitely presented groups. Suppose that G is a group and M a left $\mathbf{Z}G$-module where $\mathbf{Z}G$ denotes the integral group ring of G. If

$$\mathbf{F} : \ldots \to F_3 \to F_2 \to F_1 \to F_0 \to \mathbf{Z} \to 0$$

is a free resolution of $\mathbf{Z}$ by free left $\mathbf{Z}G$-modules then the homology groups of G with coefficients in M are given by $H_*(G, M) = H_*(M \otimes_{\mathbf{Z}G} \mathbf{F})$. Similarly, if M is a left $\mathbf{Z}G$-module, then the cohomology groups with coefficients in M are given by $H^*(G, M) = H^*(Hom_{\mathbf{Z}G}(\mathbf{F}, M))$.

If G is a finitely presented group, then we ask to what extent the sequences of abelian groups $H_*(G, M)$ and $H^*(G, M)$ can be effectively determined?

The most familiar homology group of a group G is $H_1(G, \mathbf{Z}) = G/[G, G]$, the abelianization of G. If G is finitely presented then $G/[G, G]$ is a finitely generated abelian group, and an additive finite presentation for $G/[G, G]$ can be easily written down from a presentation for G. There is then a familiar algorithm for expressing uniquely $G/[G, G]$ as a direct sum of cyclic groups (Smith normal form). This solves the isomorphism problem for $H_1(G, \mathbf{Z})$ and so this group can be effectively determined. We record this as follows:

PROPOSITION 8.1. *The homology group $H_1(G, \mathbf{Z})$ can be effectively determined from a finite presentation for G.*

Next consider the case G is a finite group which may be assumed to be given by a multiplication table presentation. The The additive group of the integral group ring $\mathbf{Z}G$ is a free abelian group of finite rank. Hence a free resolution $\mathbf{F}$ as above can be found in which each F_i is a finitely generated free abelian group (for instance the bar resolution). Now if M is a finitely presented $\mathbf{Z}G$-module, it is also finitely generated as an abelian group and its structure can be completely determined. It is easy to see that all of the maps necessary to compute the homology groups up to any dimension can be effectively computed. Hence each of the finitely generated abelian groups $H_n(G, M)$ and $H^n(G, M)$ can be effectively determined in this case.

PROPOSITION 8.2. *If G is a finite group and M is a finitely presented $\mathbf{Z}G$-module, then each of the homology groups $H_n(G, M)$ and $H^n(G, M)$ can be effectively determined.*

Of course in practice one is not so much interested in computing these groups individually as in establishing general properties of the sequences $H_*(G, M)$ and $H^*(G, M)$ for a group or a collection of groups. The above gives little information about these general questions.

Finally consider the case in which G is a polycyclic-by-finite group. As indicated above, large parts of commutative algebra can be carried out effectively, in particular Hilbert's basis theorem is effective. The analogous results for polycyclic-by-finite groups can likewise be shown to be effective (see [11]). Using this theory one can show the following:

THEOREM 8.3 ([11]). *Let G be a polycyclic-by-finite group and let M be a finitely presented $\mathbf{Z}G$-module. Then each of the homology groups $H_n(G, M)$ is a finitely generated abelian group. Moreover, there is a recursive procedure which yields for each $n \geq 0$ a finite presentation of $H_n(G, M)$. The procedure is uniform in the given data.*

In general there is little hope of effectively computing the homology groups $H_n(G, M)$ for M an arbitrary finitely presented $\mathbf{Z}G$-module even when the group G is reasonably nice. Suppose for instance that G is a free group on finitely many generators and let Q be a quotient group of G. Then $\mathbf{Z}Q$ is a cyclic $\mathbf{Z}G$-module. But Q might have unsolvable word problem, in which case the word problem for $\mathbf{Z}Q$ as a $\mathbf{Z}G$-module is unsolvable. Similarly, one can not in general decide whether Q is the trivial group, so one cannot decide whether $\mathbf{Z}Q$ is isomorphic to $\mathbf{Z}$ as a $\mathbf{Z}G$-module.

In view of these considerations it is convenient to restrict one's attention to the case M is a trivial $\mathbf{Z}G$-module and to the case $M = \mathbf{Z}$ in particular. For simplicity we use the abbreviations $H_n G = H_n(G) = H_n(G, \mathbf{Z})$.

Since $H_1 G$ can be effectively computed, it is natural to consider $H_2 G$. Now if $G = F/R$ where F is a free group and R is a normal subgroup, then Hopf's formula for $H_2 G$ is

$$H_2(G, \mathbf{Z}) = ker\{R/[F, R] \to F/[F, F]\} = (R \cap [F, F])/[F, R].$$

The abelian group $R/[F, R]$ is generated by the images of a set of defining relations for G so if G is given by a finite presentation then H_2G is finitely generated on a set of generators no larger than the number of relations of G. Despite this, the groups H_2G can not be effectively determined.

THEOREM 8.4 (Gordon [40]). *There is no algorithm to determine of an arbitrary finitely presented group G whether or not $H_2G = 0$.*

As we shall see, this result follows easily from the sorts of constructions used to prove the Adian-Rabin Theorem. However, it should be pointed out that the property $H_2G = 0$ is definitely not a Markov property of G and so the above result is not an instance of the Adian-Rabin Theorem. This follows from the following result:

THEOREM 8.5 ([16]). *Every group G which admits a recursively enumerable presentation can be embedded in a finitely presented acyclic group Q; thus by definition $H_nQ = 0$ for $n > 0$.*

While Gordon's result is not implied by the Adian-Rabin Theorem, it does follow easily from any of the constructions used to prove it. In fact the argument shows a certain class of homological properties are not recognizable. To describe these we introduce the following definition.

DEFINITION 8.1: An abstract property P of finitely presented groups is said to be a *homological Markov property* if there are two finitely presented groups G_+ and G_- such that
(1) G_+ has the property P; and
(2) if Y is a finitely presented group such that $H_nG_- \subseteq H_nY$ for $n > 1$ then Y does not have property P.

These groups G_+ and G_- will be called the *positive* and *negative witnesses* for the homological Markov property P respectively.

Note that the property $H_2G = 0$ is an example of a homological Markov property. In terms of this definition, the arguments for the Adian-Rabin Theorem show the following result which includes Gordon's result.

THEOREM 8.6. *If P be a homological Markov property of finitely presented groups, then P is not recursively recognizable.*

PROOF: We apply the Technical Lemma used in the proof of Adian-Rabin Theorem. Let Q be a finitely presented acyclic group with unsolvable word problem. Take $K = Q * G_-$ and for any word w of Q construct L_w as in the Technical Lemma. Finally put $\pi_w = L_w * G_+$. Then if $w \neq_Q 1$ it follows from the Mayer-Viettoris sequence for homology of amalgamated free products that $H_n G_- \subseteq H_n L_w \subseteq H_n(gp(\pi_w))$ for $n > 1$. So in this case $gp(\pi_w) \notin P$. On the other hand, if $w =_Q 1$ then $L_w \cong 1$ and so $gp(\pi_w) \cong G_+$ and hence $gp(\pi_w) \in P$. Since the word problem for Q is unsolvable, it follows that Q is not recursively recognizable. This completes the proof.

To describe abelian groups on a possibly infinite set of generators we use the notation $< X \mid R >_{ab}$ where $X = \{x_1, x_2, \ldots\}$ is a set of generators and $R = \{r_1, r_2, \ldots\}$ is a set of words on X. The abelian group A presented by $< X \mid R >_{ab}$ is then the quotient of the free abelian group on X by the subgroup generated by the words in R. If X is a recursively enumerable set of symbols and R a recursively enumerable set of words in those symbols we say that $< X \mid R >_{ab}$ is an *r.e. abelian group presentation*.

In [16] Baumslag, Dyer and Miller investigated the possibilities for the whole integral homology sequence $H_n G$ for a finitely presented group. Despite the fact that one knows very little about a finitely presented group from its presentation, the sequence of integral homology groups $H_n G$ turns out to be a sequence of recursively presentable abelian groups.

THEOREM 8.7 ([16]). *If G is a recursively presented group, then the integral homology sequence $H_n G$ can be described by a recursively enumerable sequence of r.e. abelian group presentations. Moreover, if G is finitely presented, the first two terms of this sequence are finitely generated.*

Whether or not a complete converse to this statement holds has not yet been resolved. However Baumslag, Dyer and Miller [16] have shown that a wide variety of r.e. sequences of recursively presentable abelian groups can be realized as the integral homology sequence of a finitely presented group. To state their results another definition is needed. An r.e. abelian group presentation $< X \mid R >_{ab}$ is called *untangled* if R is a basis of the subgroup it generates, and otherwise *tangled*. Since subgroups of free abelian groups are free they have bases. But a given r.e. abelian group presentation may be tangled and indeed may not be effectively untangled. The situation is summarized by the following result.

LEMMA 8.8 ([16]). *Let $< X \mid R >_{ab}$ be an r.e. abelian group presentation of the abelian group A.*

(1) *if A is a torsion-free abelian group there is a recursive procedure which transforms $< X \mid R >_{ab}$ into an untangled r.e. abelian group presentation $< Y \mid S >_{ab}$ of A.*

(2) if the word problem for $< X \mid R >_{ab}$ is recursively solvable there is a recursive procedure which transforms $< X \mid R >_{ab}$ into an untangled r.e. abelian group presentation $< Y \mid S >_{ab}$ of A.

However, there exist abelian groups A having an r.e. abelian group presentation but having no untangled r.e. abelian group presentations at all.

The main result on realizing a sequence of abelian groups as the integral homology of a finitely presented group is as follows:

THEOREM 8.9. Let $A_1, A_2, A_3, \ldots$ be a sequence of abelian groups in which the first two terms are finitely generated. If the A_i's are given by an r.e. sequence of r.e. abelian group presentations each of which is untangled, then there exists a finitely presented group G whose integral homology sequence is the given sequence, that is $H_n G = A_n$ for $n > 0$.

If one is interested in constructing a finitely presented group with a specified $H_n G$ for a particular n, the restriction to untangled presentations is not necessary. However, the constructions used to build such a group lose control of the homology in adjacent dimensions.

REFERENCES

[1] S. I. Adian, *Algorithmic unsolvability of problems of recognition of certain properties of groups*, Dokl. Akad. Nauk SSSR **103**, 533-535 (1955).

[2] S. I. Adian, *The unsolvability of certain algorithmic problems in the theory of groups*, Trudy Moskov. Mat. Obsc. **6**, 231-298 (1957).

[3] S. I. Adian, *Finitely presented groups and algorithms*, Dokl. Akad. Nauk SSSR **117**, 9-12 (1957).

[4] J. Alonso, *Combings of groups*, this volume.

[5] J. Alonso, T. Brady, D. Cooper, V. Ferlini, M. Lustig, M. Mihalik, M. Shapiro, and H. Short *Notes on negatively curved groups*, MSRI preprint, 1989; to appear in the proceedings of the conference "Group theory from a geometrical viewpoint" held at Trieste, 1990.

[6] D. J. Anick, *On the homology of associative algebras*, Trans. Amer. Math. Soc. **296**, 641-659 (1986).

[7] A. V. Anisimov, *Groups languages*, Kybernetika **4**, 18-24 (1971).

[8] W. Ballmann, E. Ghys, A. Haefliger, P. de la Harpe, E. Salem, R. Strebel, et M. Troyanov, " Sur les groupes hyperboliques d'après Mikhael Gromov" (Notes of a seminar held at Berne), edited by E. Ghys and P. de la Harpe, Birkhäuser, Progress in Mathematics Series, 1990.

[9] G. Baumslag, *Finitely generated residually torsion-free nilpotent groups*, in preparation.

[10] G. Baumslag, W. W. Boone and B. H. Neumann, *Some unsolvable problems about elements and subgroups of groups*, Math. Scand. **7**, 191-201 (1959).

[11] G. Baumslag, F. B. Cannonito and C. F. Miller III, *Computable algebra and group embeddings*, Journ. of Algebra **69**, 186-212 (1981).

[12] G. Baumslag, F. B. Cannonito and C. F. Miller III, *Infinitely generated subgroups of finitely presented groups, I*, Math. Zeit. **153**, 117-134 (1977).

[13] G. Baumslag, F. B. Cannonito and C. F. Miller III, *Infinitely generated subgroups of finitely presented groups, II*, Math. Zeit. **172**, 97-105 (1980).

[14] G. Baumslag, F. B. Cannonito and C. F. Miller III, *Some recognizable properties of solvable groups*, Math. Zeit. **178**, 289-295 (1981).

[15] G. Baumslag, F. B. Cannonito, D. J. S. Robinson and D. Segal, *The algorithmic theory of polycyclic-by-finite groups*, to appear in Journal of Algebra.

[16] G. Baumslag, E. Dyer and C. F. Miller III, *On the integral homology of finitely presented groups*, Topology **22**, 27-46 (1983).

[17] G. Baumslag, S. M. Gersten, M. Shapiro and H. Short, *Automatic groups and amalgams*, to appear. (A summary appears in this volume.)

[18] G. Baumslag, D. Gildenhuys and R. Strebel, *Algorithmically insoluble problems about finitely presented soluble groups, Lie and associative algebras, I*, Journ. Pure and Appl. Algebra **39**, 53-94 (1986).

[19] G. Baumslag and J. E. Roseblade, *Subgroups of direct products of free groups*, Journ. London Math. Soc. (2) **30**,44-52 (1984).

[20] G. Baumslag and D. Solitar, *Some two-generator one-relator non-Hopfian groups*, Bull. Amer. Math. Soc. **68**, 199-201 (1962).

[21] W. W. Boone, *Certain simple unsolvable problems in group theory, I, II, III, IV, V, VI*, Nederl. Akad. Wetensch Proc. Ser. A**57**, 231-237,492-497 (1954), **58**, 252-256,571-577 (1955), **60**, 22-27,227-232 (1957).

[22] W. W. Boone *Word problems and recursively enumerable degrees of unsolvability. A sequel on finitely presented groups.*, Annals of Math. **84**,49-84 (1966).

[23] W. W. Boone and G. Higman, *An algebraic characterization of the solvability of the word problem*, J. Austral. Math. Soc. *18*, 41-53 (1974).

[24] W. W. Boone and H. Rogers Jr., *On a problem of J.H.C. Whitehead and a problem of Alonzo Church*, Math. Scand. **19**, 185-192 (1966).

[25] K. S. Brown, *The geometry of rewriting systems: a proof of the Anick–Groves–Squier theorem*, this volume.

[26] J. W. Cannon, D. B. A. Epstein, D. F. Holt, M. S. Patterson and W. P. Thurston, *Word processing and group theory*, preprint, University of Warwick, 1988.

[27] C .R .J. Clapham, *Finitely presented groups with word problems of arbitrary degrees of insolubility*, Proc. London Math. Soc. (3) **14**, 633-676 (1964).

[28] C .R .J. Clapham, *An embedding theorem for finitely generated groups*, Proc. London Math. Soc. (3) **17**, 419-430 (1967).

[29] D. J. Collins, *Representation of Turing reducibility by word and conjugacy problems in finitely presented groups*, Acta Mathematica **128**, 73-90 (1972).

[30] D. J. Collins and C. F. Miller III, *The conjugacy problem and subgroups of finite index*, Proc. London Math. Soc.ser 3 **34**, 535-556 (1977).

[31] M. Dehn, *Über unendliche diskontinuerliche Gruppen*, Math. Ann. **69**, 116-144 (1911).

[32] M. J. Dunwoody, *The accessibility of finitely presented groups*, Invent. Math. **81**, 449-457 (1985).

[33] V. H. Dyson *The word problem and residually finite groups*, Notices Amer. Math. Soc. **11**, 734 (1964).

[34] E. Formanek, *Conjugacy separability in polycyclic groups*, Journ. Algebra **42**, 1-10 (1976).

[35] A. A. Fridman, *Degrees of unsolvability of the word problem for finitely defined groups*, Izdalel'stvo "Nauk", Mpscow, 193pp (1967).

[36] S. M. Gersten, *Dehn functions and l_1-norms of finite presentations*, this volume.

[37] S. M. Gersten and H. Short , *Small cancellation theory and automatic groups*, Inventiones Math. **102**, 305-334 (1990).

[38] S. M. Gersten and H. Short , *Small cancellation theory and automatic groups, Part II*, to appear in Inventiones Math.

[39] S. M. Gersten and H. Short, *Rational subgroups of biautomatic groups*, to appear in Annals of Math.

[40] C. M. Gordon, *Some embedding theorems and undecidability questions for groups*, unpublished manuscript,1980.

[41] A. V. Gorjaga and A. S. Kirkinskii, *The decidability of the conjugacy problem cannot be transferred to finite extensions of groups*, Algebra i Logica **14**, 393-406 (1975).

[42] A. P. Goryushkin, *Imbedding of countable groups in 2-generator groups*, Mat. Zametki **16**, 231-235 (1974).

[43] M. D. Greendlinger, *Dehn's algorithm for the word problem*, Comm. Pure Appl. Math. **13**, 67-83 (1960).

[44] M. D. Greendlinger, *On Dehn's algorithms for the conjugacy and word problems with applications*, Comm. Pure Appl. Math. **13**, 641-677 (1960).

[45] M. Gromov, *Hyperbolic groups*, in "Essays on group theory", MSRI series vol. 8, edited by S. Gersten, Springer-Verlag, 1987

[46] J. R. J. Groves, *Rewriting systems and homology of groups*, to appear in Proceedings of Canberra Group Theory Conf. ed. L. G. Kovacs,in Lecture Notes in Math.(Springer).

[47] J. R. J. Groves and G. C. Smith, *Rewriting systems and soluble groups*, University of Melbourne preprint, 1989.

[48] F. Grunewald and D. Segal, *Some general algorithms. I: Arithmetic groups, II: Nilpotent groups*, Annals of Math. **112** 531-583, 585-617 (1980).

[49] F. Grunewald and D. Segal, *Decision problems concerning S−arithmetic groups*, Journ. Symbolic Logic **90**, 743-772 (1985).

[50] M. Hall Jr., "The theory of groups", Macmillan, New York, 1959.

[51] P. Hall, *Finiteness conditions for soluble groups*, Proc. London Math. Soc. III4, 419-436 (1954).

[52] P. Hall, *Embedding a group in a join of given groups*, J. Austral. Math. Soc. **17**, 434-495 (1974).

[53] G. Higman, *A finitely generated infinite simple group*, J. London Math. Soc. **26**, 61-64 (1951).

[54] G. Higman, *Subgroups of finitely presented groups*, Proc. Royal Soc. London Ser. A **262**, 455-475 (1961).
[55] G. Higman, B. H. Neumann and H. Neumann, *Embedding theorems for groups*, J. London Math. Soc. **24**, 247-254 (1949).
[56] J. Hopcroft and J. Ullman, "Introduction to automata theory, languages and computation", Addison-Wesley, Boston, 1979.
[57] A. Juhasz, *Solution of the conjugacy problem in one relator groups*, this volume.
[58] O. Kharlampovich, *A finitely presented soluble group with insoluble word problem*, Izvestia Akad. Nauk Ser. Mat. **45**, 852-873 (1981).
[59] Yu. G. Kleiman, *Identities and some algorithmic problems in groups*, Dokl. Akad. Nauk SSSR **244**, 814-818 (1979).
[60] Yu. G. Kleiman, *On some questions of the theory of varieties of groups*, Dokl. Akad. Nauk SSSR **257**, 1056-1059 (1981).
[61] Yu. G. Kleiman, *On identities in groups*, Trudy Moskov Mat. Obshch. **44**, 62-108 (1982).
[62] Yu. G. Kleiman, *Some questions of the theory of varieties of groups*, Izv. Akad. Nauk SSSR Ser. Mat. **47**,37-74 (1983).
[63] A. V. Kuznetsov, *Algorithms as operations in algebraic systems*, Izv. Akad. Nauk SSSR Ser Mat (1958).
[64] P. Le Chenadec, "Canonical forms in finitely presented algebras", Res. Notes in Theoret. Comp. Sci., Pitman (London–Boston) and Wiley(New York), 1986.
[65] R. C. Lyndon, *On Dehn's algorithm*, Math. Ann. **166**, 208-228 (1966).
[66] R. C. Lyndon and P. E. Schupp, "Combinatorial Group Theory", Springer-Verlag, Berlin-Heidleberg-New York,1977.
[67] I. G. Lysënok, *On some algorithmic properties of hyperbolic groups*, Izv. Akad. Nauk SSSR Ser. Mat. **53** No. 4 (1989); English transl. in Math. USSR Izv. **35** 145-163 (1990).
[68] K. Madlener and F. Otto, *About the descriptive power of certain classes of string-rewriting systems*, Theoret. Comp. Sci. **67** 143-172 (1989).
[69] W. Magnus, *Über diskontinuierliche Gruppen mit einer definierenden Relation (Der Freiheitsats)*, J. Reine Angew. Math. **163**, 141-165 (1930).
[70] W. Magnus, *Das Identitätsproblem für Gruppen mit einer definierenden Relation*, Math. Ann.**106**, 295-307 (1932).
[71] W. Magnus, A. Karrass and D. Solitar, "Combinatorial Group Theory", Wiley, New York, 1966.
[72] A. I. Malcev, *On isomorphic matrix representations of infinite groups*, Mat. Sb. **8**, 405-422 (1940).
[73] A. I. Malcev, *Homomorphisms onto finite groups*, Ivanov. Gos. Ped. Inst. Ucen. Zap. **18**, 49-60 (1958).
[74] J. C. C. McKinsey, *The decision problem for some classes of sentences without quantifiers*, Journ. Symbolic Logic **8**, 61-76 (1943).
[75] K. A. Mihailova, *The occurrence problem for direct products of groups*, Dokl. Akad. Nauk SSSR **119**, 1103-1105 (1958).
[76] K. A. Mihailova, *The occurrence problem for free products of groups*, Dokl. Akad. Nauk SSSR **127**, 746-748 (1959).
[77] C. F. Miller III, "On group theoretic decision problems and their classification", Annals of Math. Study **68**, Princeton University Press, Princeton, NJ, 1971.
[78] C. F. Miller III, *The word problem in quotients of a group*, in "Aspects of Effective Algebra", ed. J.N. Crossley, Proceedings of a conference at Monash University Aug. 1979, Upside Down A Book Company, Steel's Creek, (1981), 246-250.
[79] C. F. Miller III and P. E. Schupp, *Embeddings into hopfian groups*, Journ. of Algebra **17**, 171-176 (1971).
[80] A. W. Mostowski, *On the decidability of some problems in special classes of groups*, Fund. Math. **59**, 123-135 (1966).
[81] D. E. Muller and P. E. Schupp, *Groups, the theory of ends and context–free languages*, J. Comput. and Sys. Sci. **26**, 295-310 (1983).
[82] D. E. Muller and P. E. Schupp, *The theory of ends, pushdown automata, and second–order logic*, Theoret. Comput. Sci. **37**, 51-75 (1985).

[83] B. H. Neumann, *Some remarks on infinite groups*, J. London Math. Soc. **12**, 120-127 (1937).

[84] B. B. Newman, *Some results on one relator groups*, Bull. Amer. Math. Soc **74**, 568-571 (1968).

[85] G. A. Noskov, *On conjugacy in metabelian groups*, Mat. Zametki **31** 495-507 (1982).

[86] P. S. Novikov *On the algorithmic unsolvability of the problem of identity*, Dokl. Akad. Nauk SSSR **85**, 709-712 (1952).

[87] P. S. Novikov *On the algorithmic unsolvability of the word problem in group theory*, Trudy Mat. Inst. Steklov **44**, 1-143 (1955).

[88] M. O. Rabin, *Recursive unsolvability of group theoretic problems*, Annals of Math. **67**, 172-194(1958).

[89] M. O. Rabin, *Computable algebra, general theory and theory of computable fields*, Trans. Amer. Math. Soc. **95**, 341-360(1960).

[90] V. N. Remmeslennikov, *Finite approximability of metabelian groups*, Algebra and Logic **7**, 268-272 (1968).

[91] V. N. Remmeslennikov, *Conjugacy in polycyclic groups*, Algebra i Logica **8**, 712-725 (1969).

[92] J. J. Rotman, "An introduction to the theory of groups", (third edition), Allyn and Bacon,Boston, 1984.

[93] I. N. Sanov, *A property of a certain representation of a free group*, Dokl. Akad. Nauk SSSR **57**, 657-659 (1947).

[94] E. A. Scott, *A finitely presented simple group with unsolvable conjugacy problem*, Journ. Algebra **90**, 333-353 (1984).

[95] E. A. Scott, *A tour around finitely presented simple groups*, this volume.

[96] P. E. Schupp, *On Dehn's algorithm and the conjugacy problem*, Math. Ann. **178**, 119-130 (1968).

[97] P. E. Schupp, *Embeddings into simple groups*, J. London Math. Soc. **13**, 90-94 (1976).

[98] D. Segal, "Polycyclic groups ", Cambridge University Press, 1983

[99] D. Segal, *Decidable properties of polycyclic groups*, to appear.

[100] C. C. Squier, *Word problems and a homological finiteness condition for monoids*, J. Pure Appl. Algebra **49**, 201-217 (1989).

[101] V. A. Tartakovskii, *The sieve method in group theory* , Mat. Sbornik **25**, 3-50(1949).

[102] V. A. Tartakovskii, *Application of the sieve method to the solution of the word problem for certain types of groups* , Mat. Sbornik **25**, 251-274 (1949).

[103] V. A. Tartakovskii, *Solution of the word problem for groups with a k-reduced basis for k > 6*, Izv. Akad. Nauk SSSR Ser. Math. **13**, 483-494 (1949).

Charles F. Miller III
Department of Mathematics
University of Melbourne
Parkville, Vic. 3052
Australia

The Word Problem for Solvable Groups and Lie Algebras

O. KHARLAMPOVICH*

My lecture is devoted to the word problem for varieties of solvable groups and Lie algebras.

First, some notation. We denote by $\mathfrak{A}^n$ the variety of solvable groups of step $\leq n$, and by $\mathfrak{N}_c$ the variety of nilpotent groups of step $\leq c$; $\mathfrak{W}_1\mathfrak{W}_2$ is the product of the varieties $\mathfrak{W}_1$ and $\mathfrak{W}_2$, $Z\mathfrak{N}_2\mathfrak{A}$ is the variety of center-by-$\mathfrak{N}_2\mathfrak{A}$ groups, $\mathfrak{B}_p$ is the variety of groups of exponent p, and $\mathfrak{A}_p$ is the variety of abelian groups of exponent p.

Suppose we have a variety $\mathfrak{W}$ of solvable groups. There are two versions of the word problem for this variety. Let FP be the class of finitely presented groups, and $\mathfrak{W}$ a given variety. The first version deals with the word problem for groups of the class $FP \cap \mathfrak{W}$. The second version deals with the word problem for groups that are relatively finitely presented in $\mathfrak{W}$. Let $FP\mathfrak{W} = \{\, G = \langle a_1, \ldots, a_n; r_1 = 1, \ldots, r_m = 1; v(x_1, \ldots, x_k) = 1 \rangle \,\}$, where v stands for the identities of the variety $\mathfrak{W}$.

Obviously, $FP \cap \mathfrak{W} \subset FP\mathfrak{W}$; therefore, if the word problem is solvable in the class $FP\mathfrak{W}$ then it is also solvable in the class $FP \cap \mathfrak{W}$.

A few words about earlier results concerning the first version of the word problem. In 1981, in my article [1], I proved that there exists a group in $\mathfrak{A}^3$ with unsolvable word problem. This result answered the well-known question posed by P. S. Novikov and S. I. Adian about the existence of a finitely presented group with unsolvable word problem, that satisfies some non-trivial identity. Actually, the group constructed in the article belongs to the variety $\mathfrak{A}_2^2\mathfrak{A} \cap \mathfrak{N}_4\mathfrak{A}$. Then, in 1986, G. Baumslag, D. Gildenhuys and R. Strebel [2] published the proof of an analogous result for the variety $Z\mathfrak{N}_3\mathfrak{A} \cap \mathfrak{A}_p^2\mathfrak{A}$. (They also proved the unsolvability of the isomorphism problem in this variety.)

Now, for some new results on the subject.

THEOREM 1 [3]. *There exists a group $G \in FP \cap Z\mathfrak{N}_2\mathfrak{A}$ with unsolvable word problem.*

---

*Research at MSRI supported in part by NSF Grant DMS-812079-05

THEOREM 2. *There exists a group $G \in FP \cap Z\mathfrak{N}_3\mathfrak{A} \cap \mathfrak{A}_p^2\mathfrak{A} \cap \mathfrak{B}_p\mathfrak{A}$ with unsolvable word problem.*

The proof of Theorem 2 and of the above-mentioned previous results is based on an interpretation of a two-tape Minsky machine with unsolvable halting problem.

The proof of Theorem 1 is based upon M. Sapir and O. Kharlampovich's result [4] on the algorithmic unsolvability of certain systems of linear differential equations over a ring of polynomials in n variables over $\mathbf{Z}$. The exact formulation of this result will be given at the end of this lecture.

I shall now talk about the word problem for the class $FP\mathfrak{W}$. We say that the word problem is solvable for a given variety $\mathfrak{W}$ if there exists an algorithm solving the word problem for any group from $FP\mathfrak{W}$, and we say that the word problem is unsolvable in $\mathfrak{W}$ if no such algorithm exists. Moreover, in this lecture, when we talk about the unsolvability of the word problem in some variety $\mathfrak{W}$, we mean that we can construct a group $G \in FP\mathfrak{W}$ having unsolvable word problem.

THEOREM 3 [5]. *If $\mathfrak{W} \supseteq Z\mathfrak{N}_2\mathfrak{A} \cap \mathfrak{B}_p\mathfrak{A}$ then the word problem for $\mathfrak{W}$ is unsolvable.*

An essential first result about this version of the word problem was proved by Remeslennikov in 1973. He constructed [6] some groups $G_n \in FP\mathfrak{A}^n$ ($n \geq 5$) with unsolvable word problem. Later, in 1979, Kukin and Epanchinzev [7] constructed similar examples in all varieties $\mathfrak{W} \supseteq \mathfrak{A}^3$. In this joint article the authors suggested that their example is from the class $FP\mathfrak{N}_2\mathfrak{A}$. Unfortunately, this stronger result is incorrect.

THEOREM 4 [8]. *If $\mathfrak{W} \subseteq \mathfrak{N}_2\mathfrak{A}$, then the word problem for $\mathfrak{W}$ is solvable.*

In connection with Theorem 4, one might mention that the solvability of the word problem in some variety $\mathfrak{W}$ is not inherited by its subvarieties.

In 1980, R. Bieri and R. Strebel [9] proved the solvability of the word problem in the class $FP \cap \mathfrak{N}_2\mathfrak{A}$. They proved that groups from this class are residually finite. In 1982, N. Romanovsky [10] proved the solvability of the word problem in the variety $Z\mathfrak{A}^2$. In my case, groups from the class $FP\mathfrak{N}_2\mathfrak{A}$ are not residually finite. Moreover, F. Canonnito and N. Gupta [11] proved that there exist finitely generated $Z\mathfrak{A}^2$-groups with unsolvable word problem.

It is of interest to investigate the boundary between the solvability and unsolvability of the word problem inside the variety $Z\mathfrak{N}_2\mathfrak{A}$.

THEOREM 5. *The word problem is solvable in the varieties $\mathfrak{N}_2\mathfrak{N}_k \cap Z\mathfrak{N}_2\mathfrak{A}$.*

Let $\mathfrak{M}_k$ be the variety defined in $Z\mathfrak{N}_2\mathfrak{A}$ by the identity

$$[[x_1, \ldots, x_{2k+2}], [y_1, \ldots, y_{2k+2}], [z_1, \ldots, z_{2k}]] = 1.$$

THEOREM 6. *The word problem is unsolvable in the varieties $\mathfrak{M}_k$.*

There are obvious inclusions

(∗) $\cdots \subseteq \mathfrak{M}_k \subseteq Z\mathfrak{N}_2\mathfrak{A} \cap \mathfrak{N}_2\mathfrak{N}_{2k+1} \subseteq \mathfrak{M}_{k+1} \subseteq \cdots.$

From Theorems 5 and 6 follows

THEOREM 7. *There exists an infinite chain (∗) of varieties of groups inside $Z\mathfrak{N}_2\mathfrak{A}$ in which the varietes with solvable and unsolvable word problem alternate.*

Hence, there is not an exact boundary between the solvability and unsolvability of the word problem in varieties inside of $Z\mathfrak{N}_2\mathfrak{A}$.

We see that the word problem is solvable in the classes $FP \cap \mathfrak{M}_k$, but in the classes $FP\mathfrak{M}_k$ it is unsolvable. This example – the first of its kind – shows that the two versions of the word problem are different. It also shows that for the varieties $\mathfrak{M}_k$ the analog of Higman's embedding theorem is untrue.

One might point out, in this context, that Mark Sapir [12] recently proved that the two versions of the word problem are identical for any variety of semigroups or associative algebras.

If a variety $\mathfrak{W}$ has unsolvable word problem but each of its proper subvarieties has a solvable word problem, then we say that $\mathfrak{W}$ is a *minimal variety with unsolvable word problem.*

THEOREM 8. *The variety of groups $Z\mathfrak{N}_2\mathfrak{A}$ contains infinitely many minimal varieties with unsolvable word problem.*

The exact description of these varieties will be given in Theorem 10.

The Theorems 1, 4-8 have exact analogs for varieties of Lie algebras over a field k of characteristic 0. In the case of Lie algebras, one can also give a description of all the varieties in $Z\mathfrak{N}_2\mathfrak{A}$ with solvable word problem.

To formulate the next results we need some definitions. First, we give these definitions for Lie algebras. Let F be a free Lie algebra of infinite rank in the variety $Z\mathfrak{N}_2\mathfrak{A}$. Let ℓ_1, ℓ_2, ℓ_3 be some elements of F'

and u, v elements of F such that $u \equiv v \pmod{F'}$; i.e., $u - v \in F'$. Then we have $(\ell_1 u)\,\ell_2\ell_3 = (\ell_1 v)\,\ell_2\ell_3$, $\ell_1\,(\ell_2 u)\,\ell_3 = \ell_1\,(\ell_2 v)\,\ell_3$, $\ell_1\ell_2\,(\ell_3 u) = \ell_1\ell_2\,(\ell_3 v)$. It therefore makes sense to consider the elements $(\ell_1 u)\,\ell_2\ell_3$, $\ell_1\,(\ell_2 u)\,\ell_3$, $\ell_1\ell_2\,(\ell_3 u)$, where u belongs to the universal enveloping algebra $U\,(F/F')$ of F/F'. If $x_1, \ldots, x_n, \ldots$ are free generators of F/F', the universal enveloping algebra $U\,(F/F')$ is isomorphic to the ring of polynomials $k\,[x_1, \ldots, x_n, \ldots]$, where k is the ground field. Let $\bar{x}_1, \ldots, \bar{x}_n, \ldots$ be some new indeterminates, then the mapping $x_i \mapsto \bar{x}_i$ can be extended to an isomorphism: $k\,[x_1, \ldots, x_n, \ldots] \to k\,[\bar{x}_1, \ldots, \bar{x}_n, \ldots]$. We denote

$$\ell_1\ell_2\ell_3 \circ u\bar{v} = (\ell_1 u)\,(\ell_2)\,(\ell_3 v)\,,$$

where $u, v \in k\,[x_1, \ldots, x_n, \ldots]$, and $\bar{v}$ is the image of v under the indicated isomorphism.

Let k denote the ground field. We say that $\alpha \in k$ is *standard* if $\alpha \neq 0, \pm 1, 2^{\pm 1}, -\frac{1}{2} \pm \frac{\sqrt{3}}{2}i$. Let us define for every $\alpha \in k$, $\alpha \neq 0$, a polynomial $P(\alpha, x)$. For standard α, the definition is

$$P(\alpha, x) = \frac{1}{\alpha^2}\,(x - \alpha\bar{x})\,(\alpha x - \bar{x})\,(x + (1 + \alpha)\bar{x})\,(\alpha x + (1 + \alpha)\bar{x})\,.$$

In the case where α is non-standard, one of the factors of the polynomial

$$(x - \alpha\bar{x})\,(\alpha x - \bar{x})\,(x + (1 + \alpha)\bar{x})\,(\alpha x + (1 + \alpha)\bar{x})$$

occurs at least twice. For this α, $P(\alpha, x)$ is the polynomial containing only one occurrence of every factor. For example, $P(-1, x) = x\,(x + \bar{x})$. For standard α, let

$$\begin{aligned}
&\tilde{P}\,(\alpha, x_1, x_2, x_3, x_4) \\
&= \frac{1}{\alpha^2}\,(x_1 - \alpha\bar{x}_1)\,(\alpha x_2 - \bar{x}_2)\,(x_3 + (1 + \alpha)\bar{x}_3)\,(\alpha x_4 + (1 + \alpha)\bar{x}_4)\,.
\end{aligned}$$

For non-standard α, $\tilde{P}\,(\alpha, x_1, x_2, x_3, x_4)$ is to be obtained from $P(\alpha, x)$ by introducing a different variable for each factor. For example,

$$\tilde{P}\,(-1, x_1, x_2, x_3, x_4) = x_1\,(x_2 + \bar{x}_2)\,.$$

We will distinguish between two types of variables. All variables of type x, y, z, with or without indices, belong to F/F', while all variables of type ℓ_i belong to F'. The varieties defined by identities of the form $\ell_1\ell_2\ell_3 \circ$

$f(x_1, \bar{x}_1, \ldots, x_n, \bar{x}_n) = 0$ we call *enveloping varieties* and we write their identities in the simpler form: $\circ f(x_1, \bar{x}_1, \ldots, x_n, \bar{x}_n) = 0$. Let $\mathfrak{W}_\alpha$ be the variety defined inside $Z\mathfrak{N}_2\mathfrak{A}$ by the identities

$$\circ \tilde{P}(\alpha, x_1, x_2, x_3, x_4)\, \tilde{P}(\alpha, x_5, x_6, x_7, x_8) = 0,$$

$$\circ \tilde{P}(\alpha, x_1, x_2, x_3, x_4)\, (x\bar{y} - y\bar{x}) = 0,$$

$$\circ (x_1\bar{x}_2 - x_2\bar{x}_1)(x_3\bar{x}_4 - x_4\bar{x}_3) = 0.$$

THEOREM 9. *The word problem is unsolvable in every variety $\mathfrak{W}$ of Lie algebras (char $k = 0$) such that for some α ($\alpha \neq 0, -1$), $\mathfrak{W}_\alpha \subseteq \mathfrak{W} \subseteq Z\mathfrak{N}_2\mathfrak{A}$. The word problem is solvable in every subvariety of $\mathfrak{W}_\alpha$ defined by some additional enveloping identity.*

Therefore, every variety $\mathfrak{W}_\alpha$ contains some minimal variety with unsolvable word problem.

To formulate the analogous results for groups we need some new definitions. Let F be a free group of infinite rank of the variety $Z\mathfrak{N}_2\mathfrak{A}$. Then the group $[F', F', F']$ is abelian, and we use for this group an additive notation. If u and v are some elements of the group ring $\mathbf{Z}[F/F']$, it makes sense to define $[\ell_1, \ell_2, \ell_3] \circ u\bar{v} = [\ell_1^u, \ell_2, \ell_3^v]$. If $x \in F/F'$, then $[\ell_1, \ell_2^x, \ell_3] = [\ell_1, \ell_2, \ell_3] \circ x^{-1}\bar{x}^{-1}$.

Let α, β be two coprime integers, $\alpha > 0$, $\beta \neq 0$. Let $\langle x, \bar{x}; x\bar{x} = \bar{x}x \rangle$ be a free abelian group on the generators $x, \bar{x}$. Let $P(\alpha, \beta, x)$ be the polynomial from $\mathbf{Z}[\langle x, \bar{x}; x\bar{x} = \bar{x}x \rangle]$ with coefficient 1 at the highest power of x, and containing all factors of the polynomial

$$\left(x^\alpha - \bar{x}^\beta\right)\left(x^\beta - \bar{x}^\alpha\right)\left(x^\alpha - \bar{x}^{-\alpha-\beta}\right)\left(x^\beta - \bar{x}^{-\alpha-\beta}\right)$$

exactly once. Let $\tilde{P}(\alpha, \beta, x_1, x_2, x_3, x_4)$ be obtained from $P(\alpha, \beta, x)$ by introducing some new variable for each factor. Let $\mathfrak{W}_{\alpha,\beta}$ be the subvariety of the variety $Z\mathfrak{N}_2\mathfrak{A}$ defined by the identity

$$\circ \tilde{P}(\alpha, \beta, x_1, x_2, x_3, x_4)\, \tilde{P}(\alpha, \beta, x_5, x_6, x_7, x_8) = 0.$$

THEOREM 10. *The word problem is unsolvable in the varieties of groups $\mathfrak{W}_{\alpha,\beta}$. If $\alpha \neq \alpha_1$ or $\beta \neq \beta_1$, then the word problem is unsolvable in the varieties $\mathfrak{W}_{\alpha,\beta} \cap \mathfrak{W}_{\alpha_1,\beta_1}$.*

In all the cases we have considered, if we have a group with an unsolvable word problem in some variety, we can also prove an unsolvability theorem for the isomorphism problem in this variety. For those cases where the word problem is solvable for all groups of a variety $\mathfrak{W}$, one can prove that the embedding problem (in a finitely generated subgroup) is also solvable. Theorems 5-10 are to be published in [13].

The proof of Theorems 1,6,8-10 in the case of unsolvability is based on the following result. Let n be a natural number, k a field. The Weyl algebra W_n over k is an associative algebra over k with the presentation

$$W_n = \text{alg}\,(x_1,\ldots,x_n,d_1,\ldots,d_n;[x_i,x_j]=[d_i,d_j]=$$
$$0,[x_i,d_j]=0(i\neq j),[d_i,x_i]=1)\,.$$

The algebra $\mathfrak{W}_n$ acts on the ring of polynomials $k\,[x_1,\ldots,x_n]$ by the operation $*$ defined by

$$x_i * f = x_i f,$$
$$d_i * f = \frac{\partial f}{\partial x_i}.$$

PROPOSITION 1 [3]. *Let $n > 9$, char $k = 0$. There exists a matrix D with entries in W_n for which there is no algorithm that decides for any natural number p whether or not there exists a solution of the system of linear equations*

$$D * \begin{pmatrix} f_1 \\ f_2 \\ \vdots \\ f_n \end{pmatrix} = \begin{pmatrix} x_n^p \\ 0 \\ \vdots \\ 0 \end{pmatrix},$$

where $f_i \in k\,[x_1,\ldots,x_n]$.

The proof of Theorem 9 in the case of solvability is based on the next result.

PROPOSITION 2 (S. V. Kotov). *There exists an effective bound for the integral solutions of the exponential Diophantine equation*

$$\sum_{i=1}^{n} \lambda_i(x)\alpha_i^x = 0,$$

where the $\lambda_i(x)$ are polynomials in x with algebraic coefficients, and the α_i are algebraic numbers with the property that α_i/α_j ($i \neq j$) are not m-th roots of unity for any m.

REFERENCES

1. O.G. Kharlampovich, *Finitely presented solvable group with unsolvable word problem*, Izvestiya Ak. Nauk, Ser. Math. **45:4** (1981), 852–873.
2. G. Baumslag, D. Gildenhuys, and R. Strebel, *Algorithmically unsolvable problems about finitely presented solvable groups, Lie and associative algebras*, J. Pure and Applied Algebra **39:1/2** (1986), 53–94.
3. O.G. Kharlampovich, *Finitely presented solvable groups and Lie algebras with unsolvable word problem*, Math. Zametki **45:5** (1989).
4. O.G. Kharlampovich and M.V. Sapir, *The word problem in the varieties of associative and Lie algebras*, Izvestiya Vuzov. Math. **3** (1989).
5. O.G. Kharlampovich, *The word problem in the variety $Z\mathfrak{N}_2\mathfrak{A}$*, Izvestiya Vuzov. Math. **11** (1988), 21–23.
6. B.N. Remeslennikov, *An example of group finitely presented in the variety $\mathfrak{A}^n, n \geq 5$, with unsolvable word problem*, Algebra i Logika **12:5** (1973), 577–602.
7. V.I. Epanchinzev and G.P. Kukin, *The word problem in the variety of groups, contained $\mathfrak{N}_2\mathfrak{A}$*, Algebra i Logika **18:3** (1979), 258–285.
8. O.G. Kharlampovich, *The word problem for subvarieties of the variety $\mathfrak{N}_2\mathfrak{A}$*, Algebra i Logika **26:4** (1987), 481-501.
9. R. Bieri and R. Strebel, *Valuations and finitely presented metabelian groups*, Proc. London Math. Soc. (Ser. III) **41:3** (1980), 439-464.
10. H.C. Romanovsky, *On the word problem for central metabelian groups*, Sib. Math. Jour. **23:4** (1982), 201–205.
11. F.B. Cannonito and N. Gupta, *On varietal analogs of Higman's embedding theorem*, Contemp. Math. **33** (1984), 134–140.
12. M.V. Sapir, *Algorithmical problems in the varieties of semigroups*, Izvestiya Vuzov. Math. **12 (1985)**, 71–74.
13. O.G. Kharlampovich, *The word problem for solvable groups and Lie algebras*, Math Sbornic. (1989) (to appear).

Ural State University, 51-Lenina stz., Sverdlovsk, USSR, 620083
Mathematical Sciences Research Institute, Berkeley, CA 94720

Solution of the Conjugacy Problem in One-Relator Groups

ARYE JUHASZ

Introduction

In this talk I sketch a proof of the following theorem:

THEOREM. *One-relator groups have solvable conjugacy problems.*

The main tool used in the proof is small cancellation theory. Let $G = \langle X|R \rangle$. If R satisfies the usual small cancellation conditions, then it is well-known that G has a solvable conjugacy problem (see [3]). In the proof of the above theorem we show that most of the one-relator groups do satisfy one of these conditions and the remaining cases can be covered using one of the following two ideas:

(1) a more flexible small cancellation condition which we call condition $W(6)$;

(2) choosing another presentation of the group, which satisfies a suitable condition on maps.

1. The structure of van Kampen diagrams for one-relator groups

1.1 Let $G = \langle X|R \rangle$ be a one-relator group, with R cyclically reduced. We classify one-relator groups according to the combinatorial properties of the cyclic word R. In particular, it will follow from this classification which one-relator groups satisfy the usual small cancellation conditions, which are the groups that satisfy $W(6)$ and which are the groups that satisfy none of them. We carry out this classification by introducing a parameter $n(R)$ for each R, as follows:

Let R be a cyclic word. Recall that a *piece* P of R is a proper subword of R such that either P occurs at least twice in R or there is also an occurrence of P^{-1} in R (see [3]). For every word R let $n(R)$ be the maximal number of occurrences of a piece of maximal length or its inverse in R. Denote this piece by A. For example, if $R = x^{-1}y^{-1}xy$ then $n(R) = 2$ and we may choose x or y to be the pieces of maximal length. If $R =$

69

$xyzuxyzxz^{-1}y^{-1}x^{-1}y$ then $n(R) = 3$ and $A = xyz$ is the (only) piece of maximal length.

If $n(R) = 0$ then either R itself occurs more than once in which case $R = U^m$ for some cyclically reduced word U, $m \geq 2$ or no proper subword of R is a piece (for example $R = xy$). In the first case the conjugacy problem has been solved by B.B. Newman [4] (see also [6], for $m \geq 3$), so we shall exclude it. In the second case, every letter occurs in R at most once, hence G is a free group. We shall exclude this class too.

By its definition, $n(R) \neq 1$. Thus we may assume $n(R) \geq 2$.

THEOREM 1. Let $G = \langle X|R \rangle$. If $n(R) \geq 6$ then G satisfies the condition $C(6)$.

The proof of the theorem is roughly this:

Let $A_1, \ldots, A_6$ be the occurrences of $A^{\pm 1}$ in R. If no two of the A_i overlap, then clearly $|R| > 6|A|$, hence every decomposition $R \equiv P_1, \ldots, P_m$ of R to pieces P_i contains at least 7 pieces. If there are A_i which overlap, we show that the sum of the length of the overlaps do not exceed $|A|$. Consequently $m|A| \geq |P_1 \ldots P_m| = |R| > 5|A|$, hence $m > 5$, i.e., $m \geq 6$ and $C(6)$ holds. (Of course the above argument holds for any value of n, not only $n = 6$.)

1.2 The $W(6)$ Theory. Recall that if a group G has a presentation which satisfies the condition $C(6)$, then the corresponding van Kampen diagrams have the property that every inner region of them has at least six neighbours. Similarly, the condition $C(4)$ and $T(4)$ implies that every inner region has at least four neighbours and no inner vertex has valency 3. The condition $C(6)$ corresponds in an obvious way to the regular tessellation of the plane by hexagons and the conditions $C(4)$ and $T(4)$ correspond to the regular tessellation of the plane by squares (see [2]). The condition $W(6)$ applies to maps which in some places look like the hexagonal tessellation (see Fig. 1) and in other places like the tessellation of the plane by squares (see Fig. 2). And, in some more places look like the following two tessellations by pentagons (see Figs. 3a, 3b).

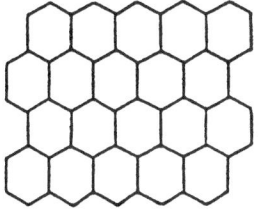

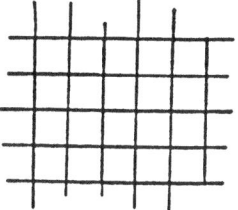

Figure 1   Figure 2

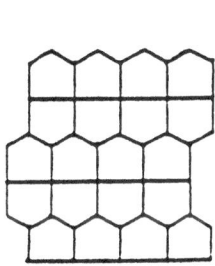

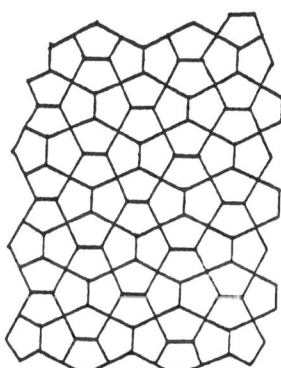

Figure 3a   Figure 3b

More precisely, given a map M, we say that M satisfies the condition $W(6)$ if for every inner region D one of the following alternatives holds:

(a) D has 4 neighbours, and no vertices with valency 3;

(b) D has 5 neighbours, and at most 3 vertices with valency 3;

(c) D has 6 neighbours, at least.

(The motivation for this definition comes from the fact that each inner region has a nonnegative hyperbolic area. The reason for not including all maps with inner regions having a nonnegative hyperbolic area is that in "practice" (see below) we can show only that certain vertices cannot have valency 3. Under this limitation the nonnegative hyperbolic area condtion boils down to the $W(6)$ condition.)

THEOREM 2. *Let* $G = \langle X | R \rangle$. *If* $n(R) \geq 4$, *then* G *satisfies* $W(6)$.

To prove the theorem we note that in a way the $W(6)$ condition says the following: If an inner region has not enough edges (i.e. less than 6) then we can compensate by showing that certain vertices have valency 4 or more. Thus the proof consists of showing that certain vertices cannot have valency 3. For the case $n = 5$ we show roughly that if $R \equiv P_1 \ldots P_5$ is a piece decomposition, then P_i is not a maximal piece unless P_i is an occurrence of $A^{\pm 1}$. Now let $R^\varepsilon \equiv HP_1^{-1}T, \varepsilon \in \{1, -1\}$. $H = H_o h$, $T = tT_o$, h, t letters. If P_1 is not a maximal piece then either P_5 ends with t^{-1} or P_2 begins with h^{-1}. Assume P_5 ends with t^{-1} and let v be the vertex between P_5 and P_1 (see Fig. 4).

We claim that v cannot have valency 3; for if $d_M(v) = 3$ then there is a region E containing v on its boundary and containing as labels the last letter (t^{-1}) of T^{-1} and the first letter (t) of P_5^{-1} as consecutive letters. But then the boundary label of E is not reduced, a contradiction.

A similar argument works for the cases $n = 4$ and $n = 3$. Thus for these cases the only piece decompositions that need special attention are those which contain occurrences of $A^{\pm 1}$. A closer analysis shows that for $n = 4$ the corresponding diagrams still satisfy $W(6)$. This, however, is not the case for $n = 2, 3$.

1.3   The cases $n = 2$ and $n = 3$.   For these cases our strategy is to glue regions along the occurrences of $A^{\pm 1}$, which in algebraic terms means that we choose a new presentation for G. Let me explain this for the case $R = A^2 Z_1 A Z_2$.

Let F be a free group with generators a, z_1, z_2. Let N be the normal subgroup of F generated by $a^2 z_1 a z_2$. For any $W \in N$ substituting $a \to A$, $z_i \to Z_i$ we obtain a relation in G. Adding them all, we obtain a new

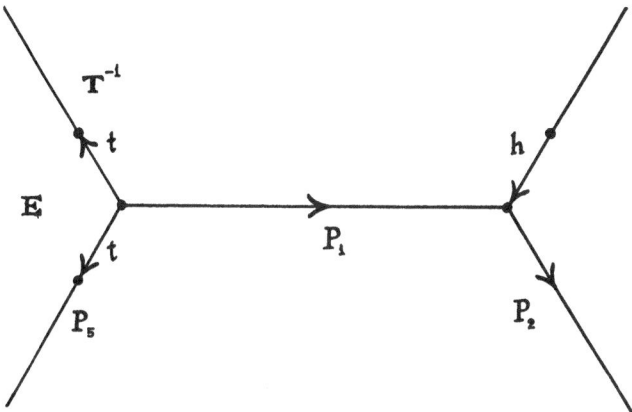

Figure 4

presentation of G. Geometrically this means that we replace our map by a new map whose regions are a union of the original regions. Following E. Rips [7] we call this a derived map.

THEOREM 3. *Assume $n = 3$ and let M be an R-diagram. Then M has a derived diagram M' which satisfies the condition $W(6)$ and one of the following holds:*

(a)(i) *Every derived region Δ which contains at least 6 original regions has the property that the length of 5 consecutive pieces does not exceed $\frac{1}{2}|\partial\Delta|$, and*

  (ii) *If L is a one layer derived diagram and Δ_1 and Δ_2 are consecutive derived regions then $|\partial\Delta_1 \cap \partial\Delta_2| \leq |R|$.*

(b) $A = B^\alpha$ *and* $R = Z^{-1}B^\alpha Z B^{-\beta}$, $\beta \geq \alpha \geq 1$, $\alpha + \beta \geq 3$.

THEOREM 4. *Assume R is one of the groups given in part (b) of Theorem 3 and let M be an annular R-diagram. Then M can be split into three annular diagrams M_1, M_2, and M_3 such that each of the following holds:*

(1) $\omega(M_1) = \omega(M)$ *and* $\tau(M_3) = \tau(M)$;

(2) M_1 *and* M_3 *satisfy the conclusions of Theorem 3(a);*

(3) M_2 *is a diagram over* $\langle B, Z \rangle$;

(4) $|\omega(M_2)| \leq 6|R|\,|\omega(M_1)|$, $|\tau(M_2)| \leq 6|R|\,|\tau(M_3)|$;

(5) M_2 has a derived map M_2' such that its dual $(M_2')*$ is an "annular tree", i.e., a one complex homotopic to a circle. Here $\omega(M)$ is the outer boundary and $\tau(M)$ is the inner boundary of M.

Finally we turn to the case $n = 2$.

For this case we introduce a new parameter, m, the maximal number of occurrences of a piece K of second to maximal length. Just as in the cases $n \geq 3$, it turns out that all maps with piece decompositions which contain no occurrences of $A^{\pm 1}$ and $K^{\pm 1}$ do satisfy the condition $W(6)$. Applying the same strategy as for $n = 3$ we get

THEOREM 5. *Assume $n = 2$ and let M be an R-diagram. Then either part (a) of Theorem 3 holds, or $R = AZ_1A^{-1}Z_2$, $Z_1 = U^\alpha, Z_2 = U^\beta$, $1 \leq \alpha \leq \beta$ and the conclusion of Theorem 4 holds for M.*

2. Solution of the conjugacy problem

In [5] P. Schupp solved the conjugacy problem for groups G having a finite $C(6)$ presentation $\langle X|R\rangle$. Denote by L the length of the longest relator in R. His solution consists of showing that two cyclically reduced elements u and v of $\langle X|-\rangle$ which have different images in G are conjugate in G if and only if they can be conjugated step-by-step by a sequence of elements $h_1, \ldots, h_r$ of length not exceeding L such that in each step the resulting element has length not exceeding $k(|u| + |v|)$, for certain constant k, depending on R. Esentially, this is the solution of the conjugacy problem for one-relator groups, except for the groups listed in Theorem 3(b) and Theorem 5. For a more precise formulation of our result, we introduce some definitions: We say that *Schupp's Procedure solves the conjugacy problem in G*, or briefly, *the conjugacy problem in G is S-solvable* if for every pair of cyclically reduced elements u and v in G, $u \underset{G}{\neq} v$, they are conjugate if and only if their conjugacy can be realized by Schupp's step-by-step bounded conjugacy procedure. We also say that u and v are *S-conjugate*. Similarly, we say that an annular diagram A over R has *property S_L*, L being a natural number if it satisfies conditions (B) below:

Denote by $\omega(A)$ and $\tau(A)$ the outer and inner boundaries of A respectively.

(B) A contains a sequence of simple closed curves $\lambda_{-a}, \lambda_{(a-1)}, \ldots, \lambda_o, \lambda_1, \ldots, \lambda_b$ homotopic to $\omega(A)$ such that

(1) $\lambda_{-a} = \omega(A)$,

(2) $\lambda_b = \tau(A)$,

(3) $|\lambda_i| \leq L(|\omega(A)| + |\tau(A)|)$,

(4) for each i, $1 \leq i \leq b$ and for each i, $-a < i \leq -1$, there exists a simple path η_i satisfying

$$o(\eta_i) \in \lambda_{i+1}, \ t(\eta_i) \in \lambda_i, \ |\eta_i| < L.$$

Thus, taking $L =$ length of the longest relator in R, we see that if every annular R-diagram has property S_L then the conjugacy problem in G is S-solvable.

In [1] we solved the conjugacy problem for $W(6)$-groups. In fact we have shown that just as in $C(6)$-groups, taking $L =$ length of the longest relator in R, $W(6)$-annular diagrams have the property S_L. This follows from the following Theorem.

THEOREM. *Let A be an annular $W(6)$ diagram. Then A splits into one-layer annular submaps $L_{-k}, L_{-(k-1)}, \ldots, L_{-1}, L_o, L_1, \ldots, L_\ell$ (L_t not necessarily with connected interior for $t \neq 0$) such that*

(1) $\omega(L_\ell) = \omega(A)$, $\tau(L_{-k}) = \tau(A)$;

(2) $|\omega(L_i)| \leq L|\omega(A)|$ $i = 0, 1, \ldots, \ell$, and $|\tau(L_j)| \leq L|\tau(A)|$, $j = 0, \ldots, k$;

(3) *If $D \in L_i$, then the number of regions in $\bigcup_{j=0}^{i} L_j$ different from D which have a common edge with D does not exceed 5.*

Combining Theorems 1 and 2 with the solution of the conjugacy problem for $C(6)$ and $W(6)$ groups, we obtain that if $n = n(R) \geq 4$ then every annular R-diagram has property S_L with $L = |R|$ hence the conjugacy problem is S-solvable.

Turning to the cases $n = 2$ and $n = 3$ we note that by Theorems 3 and 5 every R-diagram has a derived diagram which satisfies the condition $W(6)$. This however, does not solve the conjugacy problem for G without some further conditions, like those given below, since the length of the boundary label of a derived region is not bounded.

PROPOSITION. *Let $G - \langle X|R \rangle$ be a finite presentation and let L be the length of the longest relator in R. Assume that every annular R-diagram*

A has a derived diagram A' which satisfies the condition $W(6)$ and the conditions (H_1) and (H_2) below. Then

(a) A has property S_{cL} for the c of (H_2);

(b) The conjugacy problem is S-solvable for G.

(c) (H_1) There is a constant d depending recursively on $|R|$ alone, such that if the derived region Δ contains more than d original regions then the length of 5 consecutive pieces in A' does not exceed $\frac{1}{2}|\partial\Delta|$.

(d) (H_2) If M_0' is a one-layer annular diagram then there is a constant c depending recursively on $|R|$ alone such that M_0 contains two consecutive regions Δ_1 and Δ_2 having a common boundary μ with $|\mu| < cL$.

It follows from Theorems 3 and 5 and the last proposition that in all one-relator groups, except those appearing in Theorem 3(b) and Theorem 5, the conjugacy problem is S-solvable. So we shall focus on these exceptional cases.

Assume that G is one of the exceptional groups and U and V are two cyclically reduced words, $U \underset{G}{\neq} V$, which are conjugate, but are not S-conjugate. Then by Theorem 4, U is S-conjugate to U_1, V is S-conjugate to V_1, $|U| \leq 6|R||U|$, $|V_1| \leq 6|R||V|$, and U_1 and V_1 are elements over $\langle B, Z \rangle$ in case (b) of Theorem 3 and a similar assertion holds for the groups of Theorem 5. Moreover, U_1 and V_1 are conjugate over those groups. More precisely, we have the following result.

PROPOSITION. Let $G = \langle X|R \rangle$ be a one-relator group for which the conjugacy problem is not S-solvable. Then

(a) $n(R) \leq 3$;

(b) R has one of the forms described in Theorem 3(b) and Theorem 5;

(c) If W and V are cyclically reduced different (in G) conjugate elements of G which are not S-conjugate, then W is S-conjugate to W_1, V is S-conjugate to V_1, $|W_1| \leq 6|R||W|$, $|V_1| \leq 6|R||V|$ W_1 and V_1 belong to one of the following groups and are conjugate in them:

$$H_{\alpha,\beta} = \langle B, Z|B^{-\beta}ZB^\alpha Z^{-1} \rangle \ \alpha \geq 1, \beta \geq \alpha, (R = B^{-\beta}ZB^\alpha Z^{-1}, n = 3)$$

$$H_{\alpha,\beta}^1 = \langle A, U|U^{-\beta}AU^\alpha A^{-1} \rangle \ \alpha \geq 1, \beta \geq \alpha, (R = U^{-\beta}AU^\alpha A^{-1}, n = 2)$$

$$H_{(\alpha,\beta,\gamma)}^{(1)} = \langle X, Y|(XY)^{-\gamma}(YX)^\alpha(XY)^\gamma(YX)^{-\beta} \rangle,$$

$$\alpha, \gamma \geq 1, \beta \geq \alpha \ (U = \dot{Y}X, A = (XY)^\gamma, n = 2$$

$$H^{(2)}_{(\alpha,\beta,\gamma)} = \langle X, Y | (Y^{-1}X^{-\gamma}Y)X^{\alpha}(Y^{-1}X^{\gamma}Y)X^{-\beta} \rangle,$$

$$\alpha, \gamma \geq 1, \ \beta \geq \alpha \ (U = X, \ A = Y^{-1}X^{\gamma}Y, \ n = 2)$$

$$H^{(3)}_{(\alpha,\beta,\gamma)} = \langle X, Y, T | T^{-1}(XY)^{-\gamma}T(YX)^{-\alpha}T^{-1}(XY)^{\gamma}T(YX)^{-\beta} \rangle,$$

$$\alpha, \gamma \geq 1, \ \beta \geq \alpha \ (U = YX, \ A = T^{-1}(XY)^{\gamma}T, \ n = 2).$$

Thus, it only remains to solve the conjugacy problem for these particular groups $H_{\alpha,\beta}$, $H^{(i)}_{(\alpha,\beta,\gamma)}$, $i = 1, 2, 3$. This is given in the following lemma.

LEMMA. *The following groups have solvable conjugacy problems:*

(i) $G_{\alpha,\beta} \ \ = \langle x, y | y^{-1}x^{\alpha}yx^{-\beta} \rangle \, \alpha \geq 1, \ \beta \geq \alpha$

(ii) $G^{(1)}_{(\alpha,\beta,\gamma)} = \langle x, y | (xy)^{-\gamma}(yx)^{\alpha}(xy)^{\gamma}(yx)^{-\beta} \rangle, \ \alpha, \gamma \geq 1, \ \beta \geq \alpha$

(iii) $G^{(2)}_{(\alpha,\beta,\gamma)} = \langle x, y | (y^{-1}x^{-\gamma}y)x^{\alpha}(y^{-1}x^{\gamma}y)x^{-\beta} \rangle, \ \alpha, \gamma \geq 1, \ \beta \geq \alpha$

(iv) $G^{(3)}_{(\alpha,\beta,\gamma)} = \langle x, y, z | z^{-1}(xy)^{-\gamma}z(yx)^{\alpha}z^{-1}(xy)^{\gamma}z(yx)^{-\beta} \rangle, \alpha, \gamma \geq 1, \beta \geq \alpha.$

PROOF. We solve here the conjugacy problem for the groups in (i). The other cases are similar.

Let U, V be two cyclically reduced words over $\langle x, y | - \rangle$, $U \neq V$ in $G = G_{\alpha,\beta}$ and assume that U and V are conjugate in G. Then there exists an annular diagram A over R having U and V as boundary labels. Here R is the symmetrical closure of the defining relator of G. Define the derived diagram A' of A as follows; two regions D_1 and D_2 are weakly equivalent if $\partial \Delta_1 \cap \partial \Delta_2$ has label $y^{\pm 1}$. The transitive closure of weak equivalence is an equivalence relation "$\sim$". Let $\sigma(D) = \{E \in Reg(A) | E \sim A\}$ and let $\Delta(D) = Int(\bigcup_{E \in \sigma(D)} \overline{E})$. Here $Reg(A)$ stands for the set of regions of A. Then either $\Delta(D)$ is simply connected or $\partial \Delta(D)$ is homotopic to ∂A. The derived diagram A' is the diagram having regions $\Delta(D)$. If follows from this definition that the dual $(A')^*$ of A' is an "annular tree" i.e., a one-complex, homotopic to a circle. Our first task is to cut off the tree edges of $(A')^*$. For this end we introduce the elements R_n: $y^{-n}x^{\alpha^n}y^n x^{-\beta^n}$. It is easy to check that $R_n = 1$ in the group $G_{\alpha,\beta}$. An (α, β) *syllable-reduction through* $R_n^{\pm 1}$ is a replacement of $(y^{-n}x^{\alpha^n}y^n)^{\pm 1}$ by $x^{\pm\beta^n}$. Similarly, a (β, α) *syllable-reduction through* $R_n^{\pm 1}$ is a replacement of $(y^n x^{-\beta^n} n_y^{-1})^{\pm 1}$ by $x^{\pm\alpha^n}$.

Carrying out all possible syllable reductions on U and V which reduce their syllable length and applying free cancellation where possible, we get words U_1 and V_1, $U_1 \underset{G}{=} U$, $V_1 \underset{G}{=} V$ such that for a conjugating diagram A_1 corresponding to (U_1, V_1) we get that $(A_1')^*$ is a circle. Let $\Delta_1, \ldots, \Delta_t$ be the regions of A_1' such that Δ_i and $\Delta_{\overline{i+1}}$ are neighbours, $1 \le i \le t$, $0 < \overline{i+1} \le t$, $\overline{i+1} \equiv i+1 \bmod t$. Denote by μ_i the common boundary of Δ_i and $\Delta_{\overline{i+1}}$. Since $(A_1')^*$ is a circle, μ_i has label x^{ℓ_i}, $\ell_i \in \mathbf{Z}$. Let $\ell = |U_1| + |V_1|$. We propose to show that $|\ell_i| < \ell\beta^\ell$. Then, if U_1 and V_1 are conjugate in G_1, then a cyclic conjugate of U_1 is conjugate to a cyclic conjugate of V_1 by x^a, $|a| \le \ell\beta^\ell$.

Let d_i be the number of the original regions in Δ_i. Then Δ_i has a boundary label $(y^{-1}x^{d_i\alpha}yx^{-d_i\beta})^{\pm 1}$. Call the boundary path of Δ_i having label $(x^{-d_i\beta})^{\pm 1}$ the β-side of Δ_i and call the boundary path of Δ_i which has label $(x^{d_i\alpha})^{\pm 1}$ and is disjoint from the β-side of Δ_i the α-side of Δ_i. Say that μ_i is of type (u_i, v_i), $u_i, v_i \in \{\alpha, \beta\}$ if μ_i is on the u_i-side of Δ_i and on the v_i-side of $\Delta_{\overline{i+1}}$. Assume μ_i is of type (α, β), μ_i coincides with the α-side of Δ_i and is contained in the β-side of Δ_{i+1}.

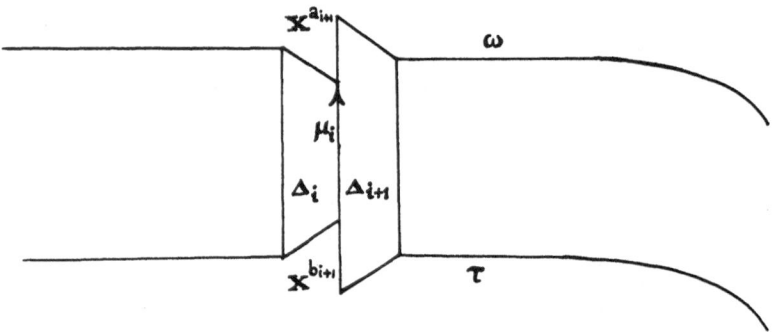

Figure 5

Then $d_{i+1}\beta = a_{i+1} + b_{i+1} + d_i\alpha$. Here a_{i+1} is the length of the head of the β-side of Δ_{i+1} which is on the outer boundary of A_1 and b_{i+1} is the length of the tail of the β-side of Δ_{i+1} which is on the inner boundary of A_1. Thus in this situation we have $d_{i+1}\beta - d_i\alpha = a_{i+1} + b_{i+1}$. In the general situation, if μ_i is of type (u_i, v_i) then we have

$$(i) \qquad\qquad u_i d_i - v_i d_{\overline{i+1}} = \varepsilon_{i+1}(a_{\overline{i+1}} + \delta b_{\overline{i+1}})$$

for suitably chosen $\varepsilon_{\overline{i+1}}, \delta_{\overline{i+1}} \in \{1, -1\}$.

It easily follows from these t equations $(1), \ldots, (t)$ that $d_i \leq \ell\beta^{t-1}$. Consequently, $|\ell_i| \leq \beta d_i \leq \ell \cdot \beta^t < \ell\beta^\ell$, as required.

Finally we describe the solution of the conjugacy problem for the general case.

Let $G = \langle X|R \rangle$ be a torsion-free one-relator group and let W and V be cyclically reduced words in $\langle X|-\rangle$, W not a cyclic conjugate of V and $W \neq V$ in G. To decide whether W is conjugate to V we do the following (we make no distinction between W and its image in G).

(1) Check whether W is S-conjugate to V. If not proceed as follows:

(2) Compute $n(R)$ and find a corresponding word A such that $A^{\pm 1}$ occurs $n(R)$ times in R.

(3) If $n(R) \geq 4$ then W and V are not conjugate.

(4) If $n(R) = 3$ and $R \neq ZU^\alpha Z^{-1}U^{-\beta}$, $\alpha \geq 1$, $\beta \geq \alpha$ then W and V are not conjugate. Here "$\neq$" means "is not a cyclic conjugate of".

(5) If $n(R) = 3$ and $R \approx ZU^\alpha Z^{-1}U^{-\beta}, \alpha \geq 1, \beta \geq \alpha$ then compute $C_{\langle Z,U\rangle}(W)$ and $C_{\langle Z,U\rangle}(V)$. Here $C_{\langle Z,U\rangle}(W)$ is the set of all the S-conjugates of W which belong to $\langle Z,U\rangle$ and are bounded by $6|R|\,|W|$. $C_{\langle Z,U\rangle}(V)$ is defined similarly. If one of these sets is empty then W and V are not conjugate. If both sets are non-empty, choose $W_1 \in C_{\langle Z,U\rangle}(W)$ and $V_1 \in C_{\langle Z,U\rangle}(V)$. Apply the last lemma to decide whether V_1 and W_1 are conjugate. W is conjugate to V if and only if W_1 is conjugate to V_1.

(6) If $n(R) = 2$ and R is not one of the exceptional words $R_{\alpha,\beta}$, $R^{(1)}_{(\alpha,\beta,\gamma)}$, $R^{(2)}_{(\alpha,\beta,\gamma)}$, $R^{(3)}_{(\alpha,\beta,\gamma)}$ then W and V are not conjugate. If R is one of the exceptional words, then proceed as in 5, replacing, of course, $R_{\alpha,\beta}$ by the corresponding exceptional words for $n = 2$.

ACKNOWLEDGEMENTS: I am grateful to Dr. E. Rips for his suggestions regarding the organization of this paper.

The basic ideas of the paper were formed while being a SERC visitor in the Mathematics Department at the University of Glasgow during Fall

1986. I am indebted to the Foundation for the grant, to Professor S. Pride for the invitation, and to the Department for its generous hospitality.

The first version of this proof was written while I was at the Weizmann Institute. I am grateful to Professor A. Joseph for his constant attention and encouragement.

REFERENCES

1. A. Juhasz, *A unified small cancellation theory I*, Journal of the London Mat. Soc. (to appear).
2. R.C. Lyndon, *On Dehn's algorithm*, Math. Ann. **166** (1966), 208–228.
3. R.C. Lyndon and P.E. Schupp, "Combinatorial Group Theory," Springer-Verlag, Berlin-New York, 1977.
4. B.B. Newman, "Aspects of One-Relator Groups," Thesis, University of Queensland, 1968.
5. P. Schupp, *On Dehn's algorithm and the conjugacy problem*, Math. Ann. **178** (1966), 119–130.
6. S. Pride, *Small cancellation conditions satisfied by one-relator groups*, Math. Z. **184** (1983), 283–286.
7. E. Rips, *Generalized small cancellation theory and application I, The word problem*, Israel J. Math. **41** (1982), p. 1-146.

Dept. of Mathematics
Technion - Israel Institute of Technology
Haifa 32000 Israel

A Tour Around Finitely Presented Infinite Simple Groups

E. A. Scott

There is a long history of mathematical interest in simple groups. Using the Higman, Neumann, and Neumann construction [3] it is easy to construct examples of infinite simple groups. For example, let C_0 be any non-trivial torsion free group. Then, by [3], there exists a torsion free group C_0^*, containing C_0, in which the non-trivial elements of C_0 are all conjugate to each other. For $i \in \mathbf{Z}$ define $C_{i+1} = C_i^*$ and let $K = \bigcup_i C_i$. Then K is an infinite simple group.

In the late 1960s R.J. Thompson [11] constructed an example of a finitely presented infinite simple group, called $G_{2,1}^+$ in this paper, and, on the basis of this work, G. Higman [5] constructed an infinite series, $G_{n,r}^+$, of such groups where n, r are positive integers, and $n \geq 2$.

K. Brown [2] has shown that another series of infinite groups, $S'_{n,r}$, which are subgroups of the $G_{n,r}^+$, are finitely presented and simple.

E.A. Scott [8] has given some different examples of finitely presented groups, which have a group of the form $G_{n,1}^+$ as a subgroup.

These three classes contain all the finitely presented infinite simple groups that we know of. We know quite a lot about the properties of the first two classes, although we still do not know, for example, precisely when two groups $G_{n,r}^+$ and $G_{n,s}^+$ are isomorphic. The scope of the third class of groups is not yet clear, but it does contain some examples of finitely presented infinite simple groups with interesting properties. In the next section we shall describe the groups in these three classes, and the rest of the paper will review what is currently known about them.

All finitely presented simple groups have solvable word problem, but they do not all have solvable conjugacy problem, see [10]. We would like to know more of what finitely presented infinite simple groups in general are like, in particular what embedding properties do they have? Higman has shown [4] that a finitely generated group G can be embedded in a finitely presented group if and only if G can be recursively presented. It is essentially a result of Kuznetsov [6] that every finitely generated subgroup of a finitely presented infinite simple group has solvable word problem. Motivated by this and Higman's subgroup theorem for finitely presented groups we have the following conjecture:

CONJECTURE.   A finitely generated group G can be embedded in a finitely presented simple group if and only if G has solvable word problem.

At the moment we have no idea whether the conjecture is true, but there is a weaker result in this direction given by Thompson [11] in 1980:

THEOREM.   *A finitely generated group has solvable word problem if and only if it can be embedded in a finitely generated simple subgroup of a finitely presented group.*

(Boone and Higman [1] proved a version of this result in 1974, but their simple groups were not necessarily finitely generated.)

1.   Preliminary Definitions And Remarks

Let $X = \{x_1, ..., x_r\}$ and $A = \{a_1, ..., a_n\}$ be sets, where we assume throughout that $n \geq 2$. We let W denote the free semigroup freely generated by A and consider the set XW. (We assume that $X \cap W = \emptyset$.)

DEFINITIONS: A *subspace* of XW is a non-empty subset U of XW such that, for all $w \in W$ and for all $u \in U$, $uw \in U$.

A subspace, U, is *cofinite* if $|XW \backslash U|$ is finite.

A subspace, U, is *inescapable* if, given any $y \in XW$, there exists some $w \in W$ such that $yw \in U$.

Any cofinite subspace, U, is inescapable because there is a finite bound on the lengths of elements *not* in U.

For $u = x_j a_{i_1} a_{i_2} ... a_{i_m} \in W$, an *initial segment* of u is an element $x_j a_{i_1} a_{i_2} ... a_{i_q}$, where $q \leq m$. Two elements u and v are *independent* if neither is an initial segment of the other.

A subset $Z \subseteq XW$ is a *(cofinite) basis* if its elements are pairwise independent and if $ZW = \{zw \mid z \in Z, w \in W\}$ is a cofinite subspace.

If $Z = \{z_1, ..., z_m\}$ is a basis then so is

$$\{z_1, ..., z_{i-1}, z_i a_1, ..., z_i a_n, z_{i+1}, ..., z_m\},$$

which is called an *elementary expansion* of Z. An *expansion* of Z is any set which can be obtained from Z by a series of elementary expansions. Any expansion of a basis is also a basis. If Y is an expansion of Z then Z is a *contraction* of Y.

It was shown by Higman [5] that any basis is an expansion of $\{x_1, ..., x_r\}$, and that any two bases Z and Y have a common expansion. Clearly, every cofinite subspace, U, has a unique basis $Z = \{ua_i \mid u \notin U, \ ua_i \in U\}$.

An *inescapable isomorphism* is a bijection $\tau : U \rightarrow V$, where U and V are inescapable subspaces, such that

$$(uw)\tau \ = \ (u\tau)w,$$

for all $u \in U$ and $w \in W$.

A *cofinite isomorphism* is an inescapable isomorphism between cofinite subspaces.

It is easy to see that the intersection, $U \cap V$, of two subspaces is a subspace, that $U \cap V$ is inescapable if U and V are inescapable, and that $U \cap V$ is cofinite if U and V are cofinite.

To make composition of functions a group operation on inescapable isomorphisms, we have to consider maximal isomorphisms.

We say that the inescapable isomorphism $\varepsilon : Y \rightarrow Z$ is an *extension* of $\tau : U \rightarrow V$ if $U \subseteq Y$ and if $u\varepsilon = u\tau$, for all $u \in U$.

DEFINITION: A *maximal inescapable isomorphism* is an inescapable isomorphism which has no proper extensions.

It is straightforward to check that every inescapable isomorphism has a unique maximal extension, and that for cofinite isomorphisms, the maximal extension is also cofinite.

NOTATION:

$$G_{n,r} \ = \ \{\text{maximal cofinite isomorphisms}\}$$
$$\mathcal{G}_{n,r} \ = \ \{\text{maximal inescapable isomorphisms}\}$$

For maximal inescapable isomorphisms

$$\tau : U \ \rightarrow \ V \quad \text{and} \quad \varepsilon : Y \ \rightarrow \ Z,$$

we define $\tau \circ \varepsilon$ to be the maximal extension of the inescapable isomorphism given by the rule $u \rightarrow (u\tau)\varepsilon$, for all $u \in U$ such that $u\tau \in Y$.

LEMMA 1. $\mathcal{G}_{n,r}$ is a group under the operation $\circ$, and $G_{n,r}$ is a subgroup of $\mathcal{G}_{n,r}$.

DEFINITIONS: Order XW lexicographically, then:
   An element $\theta \in G_{n,r}$ is said to be *order preserving* if $u < v$ implies that $u\theta < v\theta$, whenever $u\theta, v\theta$ are defined.
   An element $\theta \in G_{n,r}$ is said to be *cyclic order preserving* if there exists some basis $u_1 < ... < u_m$ such that, for some i,

$$u_i\theta < ... < u_m\theta < u_1\theta < ... < u_{i+1}\theta.$$

NOTATION:

$$F_{n,r} = \{\theta \in G_{n,r} \mid \theta \text{ is order preserving}\}$$
$$T_{n,r} = \{\theta \in G_{n,r} \mid \theta \text{ is cyclic order preserving}\}$$

It follows from Lemma 9 that $T_{n,r}$ is a subgroup of $G_{n,r}$. In fact, $T_{n,r}$ is generated by $F_{n,r}$ together with one other element, μ. The element μ is defined precisely in section 4, but in the case where $r > n + 1$ we have $\mu(x_i w) = x_{i+1}w, \ 1 \leq i \leq n - 1$, and $\mu(x_r w) = x_1 w$, for all $w \in W$.

NOTATION: $S_{n,r} = \langle\, F_{n,r}\,,\, \mu^e\,\rangle,$ where $e = hcf(r, n - 1)$.

$S_{n,r}$ is the group that Brown [2] calls $T_{n,r}^0$.

   The derived subgroups of $G_{n,r}$ and $S_{n,r}$ are finitely presented infinite simple groups.

   It is known that any subgroup of $\mathcal{G}_{n,r}$ that contains $G_{n,r}$ will have a simple derived subgroup. For this reason we also consider certain subgroups of $\mathcal{G}_{n,r}$ that contain $G_{n,r}$. The following ideas will be defined explicitly later, but roughly speaking, an H-symbol for an element $\sigma \in \mathcal{G}_{n,r}$ is a finite expression which defines σ in terms of two cofinite bases and finitely many of the elements of the subgroup H of $\mathcal{G}_{n,1}$. A subgroup, K, of $\mathcal{G}_{n,r}$ is said to be H-expansible if every element of K has a H-symbol and if any two H-symbols for $\sigma_1, \sigma_2 \in K$ can be combined to form an H-symbol for the product $\sigma_1 \sigma_2$.

For any $H \leq \mathcal{G}_{n,1}$, if $\mathcal{H} = \langle \mathcal{G}_{n,1}, H \rangle$ is H-expansible, then we can give an explicit set of relations of $\mathcal{H}$ which, together with sets of defining relations for $G_{n,r}$ and H, form a set of defining relations for $\mathcal{H}$. Very little is known about the types of groups H for which $\mathcal{H}$ is H-expansible, but there is a class, $\mathcal{C}_n$, of subgroups of $\mathcal{G}_{n,1}$ for which it is known that $\mathcal{H}$ is H-expansible. For H in $\mathcal{C}_n$ we know that if H is finitely presented then so is $\mathcal{H}$. We cannot always guarantee that the derived subgroup $\mathcal{H}'$ is also finitely presented, but we can always choose some integer m such that H is also in $\mathcal{C}_m$ and the derived subgroup $\langle G_{m,r}, H \rangle'$ is a finitely presented simple group. We have no idea what types of groups can be embedded into finitely presented simple groups using this type of construction, although we do know that all the groups in all the $\mathcal{C}_n$ are residually finite. However, we can construct some new examples of finitely presented infinite simple groups in this way, and using a slight variation we can construct a finitely presented simple group that has an unsolvable conjugacy problem.

The finitely presented infinite simple groups that are currently known fall into the following three classes:
  (i) The derived subgroups of the $G_{n,r}$.
  (ii) The derived subgroups of the $S_{n,r}$.
  (iii) The derived subgroups of certain H-expansible groups, $\langle G_{n,r}, H \rangle$.

2.   The Groups $G_{n,r}$

In this section we will outline some results about the groups $G_{n,r}$ which were all proved by Higman [5].

THEOREM 2.   *If n is even then $G_{n,r}$ is a finitely presented infinite simple group. If n is odd then $G_{n,r}$ has a finitely presented infinite simple subgroup $G_{n,r}^+$ of index 2.*

It is worth noting here that the situation for the groups $S_{n,r}$ is slightly different. If $hcf(r, n-1) = 1$ then $S_{n,r}$ is a finitely presented simple group, otherwise the finitely presented simple group $S'_{n,r}$ has infinite index in $S_{n,r}$ and the quotient group, $S_{n,r}/S'_{n,r}$, is a free Abelian group of finite rank.

To unify notation we set $G_{n,r} = G^+_{n,r}$ when n is even.

THEOREM 3.   *If p is a prime which does not divide $n - 1$ then $G^+_{n,r}$ has exactly n conjugacy classes of elements of order p.*

In the next section we describe an example of an H-expansible group, $\mathcal{H} = \langle G_{2,1} , H \rangle$, such that $\mathcal{H}'$ is a finitely presented simple group with infinitely many conjugacy classes of elements of order 2. In fact, given any finite set, P, of primes, it is possible to construct a finitely presented simple group which has infinitely many conjugacy classes of elements of order p if and only if $p \in P$.

An Alternative View Of $G_{n,r}$:

Consider sets S on which there is defined an n-ary operation $\lambda : S^n \to S$ and n unary operations $a_i : S \to S$. Form the variety $\mathcal{V}_n$ of all universal algebras which satisfy the laws:

$$(sa_1, ...sa_n)\lambda = s$$
$$(s_1, ..., s_n)\lambda a_i = s_i, \qquad 1 \leq i \leq n.$$

Then, if $V_{n,r}$ is the free algebra of $\mathcal{V}_n$ which is freely generated by r generators, $G_{n,r}$ is isomorphic to the automorphism group of $V_{n,r}$.

THEOREM 4.
  (i) If $G_{n,r} \cong G_{m,s}$ then $n = m$.
  (ii) If $r \equiv s \bmod n - 1$, then $G_{n,r} \cong G_{n,s}$.

Part (ii) of this theorem is also true for the groups $S_{n,r}$, see Theorem 19.

The first part of Theorem 4, which shows that there exist infinitely many non-isomorphic finitely presented infinite simple groups, follows from Theorem 3. The second part can easily be seen from the alternative view of $G_{n,r}$, for, if $\{x_1, ..., x_r\}$ is a free set of generators for $V_{n,r}$ then so is $\{x_1 a_1, ..., x_1 a_n, x_2, ..., x_r\}$, and so $V_{n,r} \cong V_{n,r+n-1}$.

We can give more detailed necessary conditions and sufficient conditions for groups of the form $G_{n,r}^+$ to be isomorphic. Let P be the set of residue classes of integers modulo $n - 1$. Let Φ be the set of classes in P that are prime to $n - 1$, and let Φ_0 be the subgroup of Φ generated by the divisors of n. Then:

THEOREM 5.

(i) If r and s belong to the same orbit of P under the action of Φ_0 then
$G_{n,r}^+ \cong G_{n,s}^+$.

(ii) If r and s do not belong to the same orbit of P under the action of Φ then $G_{n,r}^+ \not\cong G_{n,s}^+$.

However, these conditions do not solve the isomorphism problem for this class of groups because, for example, they do not tell us whether or not $G_{40,1} \cong G_{40,7}$.

THEOREM 6.   The order and conjugacy problems for $G_{n,r}^+$ are solvable.

THEOREM 7.   Every countable locally finite group can be embedded in every $G_{n,r}^+$.

This result is not in general true for the groups $S_{n,r}$, for example if $n - 1$ does not divide r then $T_{n,r}$ has no elements of order $n-1$, (see Theorem 18).

In contrast to Theorem 7, the next theorem shows that there are strong restrictions on the torsion-free subgroups of the $G_{n,r}$, and hence of $S_{n,r}$.

THEOREM 8.   Let A be a torsion-free Abelian group of finite rank contained in $G = G_{n,r}$. Then A is free Abelian of finite rank and $C_G(A)$ has finite index in $N_G(A)$. Futhermore, $N_G(A)$ has a direct factor B which has a free Abelian subgroup C such that C has finite index in AB.

From Theorem 8 we can see that the group $GL(3, \mathbf{Z})$, which is a finitely generated group with solvable word problem, cannot be embedded in any

$G_{n,r}$. For, the Abelian group

$$A = \left\{ \begin{pmatrix} 1 & a & b \\ 0 & 1 & 0 \\ 0 & 0 & 1 \end{pmatrix} \;\middle|\; a, b \in \mathbf{Z} \right\}$$

is normalized but not centralized by elements of the form

$$\begin{pmatrix} 1 & a & b \\ 0 & 1 & c \\ 0 & 0 & 1 \end{pmatrix},$$

where $c \neq 0$. However, as we will see towards the end of this paper, it is possible to embed $GL(3, \mathbf{Z})$ in a finitely presented simple group that is the derived subgroup of some H-expansible group.

3.   Symbols

The results on finitely presented simple groups given in [5] and [8] were proved by manipulating symbols which represent group elements. In this section we will describe these symbols and the way in which they are combined.

If $r = 1$ we ignore X and identify $x_1 w$ with w, for all $w \in W$. Using this identification we define symbols.

An H-symbol is an object, Γ, of the form

$$\begin{pmatrix} u_1 & u_2 & \cdots & u_m \\ h_1 & h_2 & \cdots & h_m \\ v_1 & v_2 & \cdots & v_m \end{pmatrix}$$

where $h_i \in H \leq \mathcal{G}_{n,1}$, $1 \leq i \leq m$, and $\{u_1, ..., u_m\}$, $\{v_1, ..., v_m\}$ are bases in XW, $|X| = r$.

We say that $h \in \mathcal{G}_{n,r}$ has the symbol Γ if

$$(u_i w)h = v_i(w h_i),$$

for all w such that $w h_i$ is defined.

If $\begin{pmatrix} w_1 & \cdots & w_t \\ \ell_1 & \cdots & \ell_t \\ y_1 & \cdots & y_t \end{pmatrix}$ is an H-symbol for h_i, then using the identification
of $x_1 w$ with w,

$$\begin{pmatrix} u_1 & \cdots & u_{i-1} & u_i w_1 & \cdots & u_i w_t & u_{i+1} & \cdots & u_s \\ h_1 & \cdots & h_{i-1} & \ell_1 & \cdots & \ell_t & h_{i+1} & \cdots & h_s \\ v_1 & \cdots & v_{i-1} & v_i y_1 & \cdots & v_i y_t & v_{i+1} & \cdots & v_s \end{pmatrix}$$

is an H-symbol for h. So, whilst each symbol determines a unique element
of $\mathcal{G}_{n,r}$, an element of $\mathcal{G}_{n,r}$ may have many different H-symbols, for a given
H.

If $k \in \mathcal{G}_{n,r}$ and $g \in \mathcal{G}_{n,r}$ have H-symbols

$$\begin{pmatrix} u_1 & u_2 & \cdots & u_s \\ h_1 & h_2 & \cdots & h_s \\ v_1 & v_2 & \cdots & v_s \end{pmatrix} \quad \text{and} \quad \begin{pmatrix} v_1 & v_2 & \cdots & v_s \\ f_1 & f_2 & \cdots & f_s \\ y_1 & y_2 & \cdots & y_s \end{pmatrix},$$

respectively, then the product kg has H-symbol

$$\begin{pmatrix} u_1 & u_2 & \cdots & u_s \\ h_1 f_1 & h_2 f_2 & \cdots & h_s f_s \\ y_1 & y_2 & \cdots & y_s \end{pmatrix}.$$

This symbol is the *combination* of the first two.

It is easy to check that k^{-1} has H-symbol

$$\begin{pmatrix} v_1 & v_2 & \cdots & v_s \\ h_1^{-1} & h_2^{-1} & \cdots & h_s^{-1} \\ u_1 & u_2 & \cdots & u_s \end{pmatrix}.$$

DEFINITION: Let H be a subgroup of $\mathcal{G}_{n,1}$. A subgroup K of $\mathcal{G}_{n,r}$ is said to
be *H-expansible* if there exists a set $\{k_i\}_{i \in I}$ of generators for K such that
each k_i has an H-symbol, and if $k = k_{i_1}^{\delta_1} k_{i_2}^{\delta_2} \ldots k_{i_m}^{\delta_m}$ has an H-symbol which
is a combination of the H-symbols for the $k_{i_j}^{\delta_j}$, where $\delta_j = \pm 1$. (It can be
shown [8] that if K is H-expansible with respect to one set of generators,
then it is H-expansible with respect to any set of generators.)

What this definition is saying is that every element of K has an H-symbol
and every relation in K has an H-symbol which is a combination of these
H-symbols.

EXAMPLE: Let $n = 2$ and $r = 1$.   Define

$$U = \{w \mid w \in W \text{ contains a subword of the form } a_2a_1 \text{ or } a_2a_2\}.$$

Then, since U is clearly an inescapable subspace of W, the map $t' : U \to U$ given by

$$(a_1^i a_2 a_1 w)t' = a_1^i a_2 a_2 w$$
$$(a_1^i a_2 a_2 w)t' = a_1^i a_2 a_1 w, \quad \text{for all } w \in W,$$

is an inescapable isomorphism. Let $t \in \mathcal{G}_{n,1}$ be the maximal extension of t', and let $H = \langle t \rangle$, so H is a cyclic group of order 2. Then t has H-symbol

$$\begin{pmatrix} a_1 & a_2a_1 & a_2a_2 \\ t & 1 & 1 \\ a_1 & a_2a_1 & a_2a_2 \end{pmatrix}.$$

It is not hard to see that $\langle G_{2,1}, H \rangle$ is H-expansible, and the methods of [8] can be used to show that $\mathcal{H} = \langle G_{2,1}, H \rangle$ is finitely presented and that $\langle G_{2,1}, H \rangle'$ has index 2 in $\mathcal{H}$. A detailed study of the H-symbols for elements of order 2 in $\mathcal{H}$ shows that $\mathcal{H}$, and hence $\mathcal{H}'$, has infinitely many conjugacy classes of elements of order 2, and finitely many conjugacy classes of elements of order p, for any odd prime p.

We will return to symbols in general later, but for a while we will just consider *simple symbols*, those for which $h_i = 1$ for all i,

$$\begin{pmatrix} u_1 & u_2 & \cdots & u_m \\ 1 & 1 & \cdots & 1 \\ v_1 & v_2 & \cdots & v_m \end{pmatrix}.$$

Clearly, $h \in \mathcal{G}_{n,r}$ has a simple symbol if and only if $h \in G_{n,r}$. Since any two bases have a common expansion, and since $1 \in H$, it is easily seen that $G_{n,r}$ is H-expansible, for any $H \leq \mathcal{G}_{n,1}$.

It is because the rules for manipulating simple symbols are essentially finite that we can prove [5] that $G_{n,r}$ is finitely presented.

4.   Subgroups Of $G_{n,r}$

In this section we consider the groups $F_{n,r}$, $S_{n,r}$, and $T_{n,r}$ in more detail. The results here are due to Brown [2], but Brown is more interested in the homological properties of the groups he studies, and his work is written up reflecting this point of view. In order to give a unified presentation of what is currently known about finitely presented infinite simple groups, some of Brown's proofs have been re-written here, using the approach and terminology of [5] and [8].

LEMMA 9.   *Suppose that $\theta \in G_{n,r}$.*
(i) *If there exists some basis $u_1 < ... < u_m$ such that $u_1\theta < ... < u_m\theta$, then $\theta \in F_{n,r}$.*
(ii) *If $\theta \in T_{n,r}$ then for any basis $u_1 < ... < u_m$ in the domain of θ, there is some i such that $u_i\theta < ... < u_m\theta < u_1\theta < ... < u_{i+1}\theta$.*

PROOF: First note that

$$u\theta < v\theta \quad \Rightarrow \quad ua_i\theta < va_i\theta,$$

for all u, v in the domain of θ, $1 \leq i \leq n$, and if u is not of the form va_j then

$$u < va_i \Leftrightarrow u < v.$$

Suppose that $u_1 < ... < u_m$ is a basis in the domain of θ, and that there is some i such that $u_i\theta < ... < u_m\theta < u_1\theta < ... < u_{i+1}\theta$. From the first remark above, it is clear that θ preserves the cyclic order of a simple expansion of $\{u_1, ..., u_m\}$, and hence that θ preserves the cyclic order of any expansion of this basis.

Now suppose that there is some j such that $u_j = ua_1, ..., u_{j+n-1} = ua_n$, for some u in the domain of θ. Since $u\theta a_n \not< u\theta a_1$ we must have

$$u_i\theta < ... < u\theta a_1 < ... < u\theta a_n < ... < u_{i+1}\theta.$$

Thus $u_i\theta < ... < u_{j-1}\theta < u\theta < u_{j+1}\theta < ... < u_{i+1}\theta$. So θ preserves the cyclic order of a simple, and hence any, contraction of this basis, on which it is defined.

We know from [5] that any basis on which θ is defined is an expansion of some unique shortest basis in the domain of θ. Thus if θ preserves the cyclic order of some basis, it must, by contraction, preserve the cyclic order of the shortest basis, and hence, by expansion, of any basis on which it is defined. This proves (ii).

If θ satisfies the condition in (i) then we see from the above discussion that θ must preserve the order of any basis in its domain. Suppose that $u < v$, and that u, v lie in the domain of θ. To prove (i), we want to show that $u\theta < v\theta$. If u is an initial segment of v then, for some w, $v\theta = u\theta w$ and so $u\theta < v\theta$ trivially. If u is not an initial segment of v then there is some basis which contains them both. Since $u\theta$ and $v\theta$ are defined we can assume that this basis lies in the domain of θ. Then we must have $u\theta < v\theta$, as required. □

The next two theorems look at the groups $F_{n,r}$ and $S_{n,r}$ as subgroups of $T_{n,r}$. The first shows that $T_{n,r}$ is generated by $F_{n,r}$ and one other element, the second shows that $S_{n,r}$ is a normal subgroup of finite index in $T_{n,r}$.

In what follows it is convenient to have $r \geq n + 1$. However, since a large portion of the later part of this paper deals solely with the cases when $r = 1$, I do not want to assume that we are working in some $G_{n,r}$ with $r \geq n + 1$ (even though $G_{n,1} \cong G_{n,2n-1}$). Thus we define the subset $\chi_R = \{\xi_1, \ldots, \xi_R\}$ of $X_r W$ as follows:

If $r \geq n + 1$, we take $R = r$ and $\xi_i = x_i$, $1 \leq i \leq r$.

If $2 \leq r \leq n$, we take $R = n - 1 + r$ and let

$$\chi_R = \{x_1 a_1, \ldots, x_1 a_n, x_2, \ldots, x_r\}.$$

If $r = 1$, we take $R = 2n - 1$ and let

$$\chi_R = \{x_1 a_1 a_1, \ldots, x_1 a_1 a_n, x_1 a_2, \ldots, x_1 a_n\}.$$

So we always have $R \geq n + 1 \geq 3$.

In what follows μ is the element of $T_{n,r}$ that has symbol

$$\begin{pmatrix} \xi_1 & \cdots & \xi_{R-1} & \xi_R \\ 1 & \cdots & 1 & 1 \\ \xi_2 & \cdots & \xi_R & \xi_1 \end{pmatrix}.$$

THEOREM 10. *Every element of $T_{n,r}$ can be written in the form $\delta\mu^i\alpha$, for some elements $\delta, \alpha \in F_{n,r}$. So, in particular, $T_{n,r}$ is generated by $F_{n,r}$ together with the element μ.*

PROOF: First consider any element, τ, which has a symbol of the form

$$\begin{pmatrix} v_1 & v_2 & \cdots & v_m \\ 1 & 1 & \cdots & 1 \\ v_{j+1} & v_{j+2} & \cdots & v_j \end{pmatrix},$$

where $v_1 \leq \ldots \leq v_m$, $j \geq 1$, and $m \geq R$. Let i be congruent to j modulo $n-1$, with $1 \leq i \leq n-1$, let $\ell = j - i + 1$ and let $k = m - R - \ell + 1$. Then since $R \geq n + 1$, we have $i + 2 \leq R$ and we can choose a symbol for μ^i of the form

$$\begin{pmatrix} \xi_1 & \xi_2 y_1 & \cdots & \xi_2 y_k & \xi_3 & \cdots & \xi_{R-1} & \xi_R z_1 & \cdots & \xi_R z_\ell \\ 1 & 1 & \cdots & 1 & 1 & \cdots & 1 & 1 & \cdots & 1 \\ \xi_{i+1} & \xi_{i+2} y_1 & \cdots & \xi_{i+2} y_k & \xi_{i+3} & \cdots & \xi_{i-1} & \xi_i z_1 & \cdots & \xi_i z_\ell \end{pmatrix}.$$

Then let β, α be the elements of $F_{n,r}$ which have symbols

$$\begin{pmatrix} v_1 & v_2 & \cdots & v_{k+1} & v_{k+2} & \cdots & v_{m-\ell} & v_{m-\ell+1} & \cdots & v_m \\ 1 & 1 & \cdots & 1 & 1 & \cdots & 1 & 1 & \cdots & 1 \\ \xi_1 & \xi_2 y_1 & \cdots & \xi_2 y_k & \xi_3 & \cdots & \xi_{R-1} & \xi_R z_1 & \cdots & \xi_R z_\ell \end{pmatrix}$$

and

$$\begin{pmatrix} \xi_1 & \cdots & \xi_{i-1} & \xi_i z_1 & \cdots & \xi_i z_\ell & \xi_{i+1} & \xi_{i+2} y_1 & \cdots & \xi_{i+2} y_k & \cdots & \xi_R \\ 1 & \cdots & 1 & 1 & \cdots & 1 & 1 & 1 & \cdots & 1 & \cdots & 1 \\ v_1 & \cdots & v_{i-1} & v_i & \cdots & v_j & v_{j+1} & v_{j+2} & \cdots & v_{j+k+1} & \cdots & v_m \end{pmatrix}$$

respectively. Then $\tau = \beta \mu^i \alpha$.

Now, a general element, $\theta \in T_{n,r}$ has a symbol of the form

$$\begin{pmatrix} u_1 & u_2 & \cdots & u_m \\ 1 & 1 & \cdots & 1 \\ v_{j+1} & v_{j+2} & \cdots & v_j \end{pmatrix},$$

where $u_1 \leq \ldots \leq u_m$, $v_1 \leq \ldots \leq v_m$, and $m \geq R$. Thus $\theta = \psi \beta \mu^i \alpha$, where $\psi \in F_{n,r}$ has symbol

$$\begin{pmatrix} u_1 & u_2 & \cdots & u_m \\ 1 & 1 & \cdots & 1 \\ v_1 & v_2 & \cdots & v_m \end{pmatrix}.$$

$\square$

We call an element, $\tau \in G_{n,r}$, a cycle if it has a symbol of the form

$$\begin{pmatrix} v_1 & \cdots & v_{m-1} & v_m \\ 1 & \cdots & 1 & 1 \\ v_2 & \cdots & v_m & v_1 \end{pmatrix}.$$

Clearly, all cycles lie in $T_{n,r}$, only the trivial ones lie in $F_{n,r}$, and every element of $T_{n,r}$ can be written in the form $\delta\tau^i$, where $\delta \in F_{n,r}$ and τ is a cycle.

LEMMA 11. For $\theta \in G_{n,r}$, if $\theta = \delta\tau^i$ and $\theta = \gamma\varepsilon^j$, where $\delta, \gamma \in F_{n,r}$ and τ, ε are a cycles, then i and j are congruent modulo $n - 1$.

PROOF: Let

$$\begin{pmatrix} u_1 & \cdots & u_m \\ 1 & \cdots & 1 \\ v_{k+1} & \cdots & v_k \end{pmatrix}$$

be the unique shortest symbol for θ. We will show that if $\theta = \delta\tau^i$ then $i \equiv k \ (n - 1)$.

First notice that for any symbols

$$\begin{pmatrix} u_1 & \cdots & u_m \\ 1 & \cdots & 1 \\ v_{k+1} & \cdots & v_k \end{pmatrix} = \Gamma_1 \quad \text{and} \quad \begin{pmatrix} y_1 & \cdots & y_s \\ 1 & \cdots & 1 \\ z_{\ell+1} & \cdots & z_\ell \end{pmatrix} = \Gamma_2,$$

if Γ_2 is a simple, and hence any, expansion of Γ_1 then ℓ is congruent to k modulo $n - 1$.

If τ^i has symbol

$$\begin{pmatrix} w_1 & \cdots & w_m \\ 1 & \cdots & 1 \\ w_{i+1} & \cdots & w_i \end{pmatrix},$$

then there is an expansion

$$\begin{pmatrix} q_1 & \cdots & q_s \\ 1 & \cdots & 1 \\ z_{\ell+1} & \cdots & z_\ell \end{pmatrix},$$

of this symbol such that there is a symbol of the form

$$\begin{pmatrix} y_1 & \cdots & y_s \\ 1 & \cdots & 1 \\ z_{\ell+1} & \cdots & z_\ell \end{pmatrix},$$

for θ. This last symbol must be a expansion of the shortest symbol for θ, so we have $k \equiv \ell \equiv i$, as required. $\qquad\Box$

Recall that $S_{n,r} = \langle F_{n,r}, \mu^e \rangle$, where $e = hcf(r, n - 1)$.

Since, by Theorem 15, $T_{n,r}$ is finitely presented, the next result shows that $S_{n,r}$ is finitely presented.

THEOREM 12. *The group $S_{n,r}$ is the kernel of a surjective homomorphism $f : T_{n,r} \to \mathbf{Z}_e$. Futhermore, every element of $S_{n,r}$ can be written in the form $\delta\mu^{ce}\alpha$, where $\delta, \alpha \in F_{n,r}$.*

PROOF: Let $g : \mathbf{Z} \to \mathbf{Z}_e$ be the natural homomorphism, and define f by the rule $(\delta\tau^i)f = ig$. Then, by Lemma 11, f is well defined, and clearly f is surjective. Any two elements, $\theta, \varphi \in T_{n,r}$ have symbols

$$\begin{pmatrix} u_1 & \cdots & u_s \\ 1 & \cdots & 1 \\ v_{j+1} & \cdots & v_j \end{pmatrix} \quad \text{and} \quad \begin{pmatrix} v_1 & \cdots & v_s \\ 1 & \cdots & 1 \\ w_{k+1} & \cdots & w_k \end{pmatrix},$$

which can be combined. So f is a homomorphism.

Suppose that $\theta \in \ker f$, and choose a symbol

$$\begin{pmatrix} u_1 & \cdots & u_s \\ 1 & \cdots & 1 \\ v_{j+1} & \cdots & v_j \end{pmatrix},$$

for θ, where $s \geq R$. Then $j = ce$, for some integer c. If i is congruent to j modulo $n - 1$ then e divides i. So the result follows from Theorem 10. □

Next we discuss the finite presentation of the groups $G_{n,r}$, $T_{n,r}$, $S_{n,r}$, and $F_{n,r}$.

We say that an element of $G_{n,r}$ has *depth* d if it has a symbol of length $m = r + (n-1)d$. Of course, an element of depth d is also of depth $d + 1$.

LEMMA 13.   *The groups $G_{n,r}$, $T_{n,r}$, $S_{n,r}$, and $F_{n,r}$ are each generated by their elements of depth three.*

PROOF: The result for $G_{n,r}$ was proved in [5]. We will prove the result for $F_{n,r}$, then, since μ and μ^e have depth 3, the results for $T_{n,r}$ and $S_{n,r}$ will follow from this.

Suppose that $\theta \in F_{n,r}$ has depth $d \geq 4$. We will show that θ is a product of elements of $F_{n,r}$ of depth at most $d - 1$, then the result follows by induction.

Choose an ordered basis

$$u_1 < \dots < u_i < u_{i+1} = ua_1 < \dots < u_{i+n} = ua_n < u_{i+n+1} < \dots < u_m$$

in the domain of θ, where $m = r + (n-1)d$. If $v_j = u_j\theta$, $1 \le j \le m$, then, since $\theta \in F_{n,r}$, $v_1 < \dots < v_m$. For some j and some v, $v_{j+k} = va_k$, $1 \le k \le n$. Taking θ^{-1} if necessary, we may assume that $i \ge j$.

We consider two cases:

$j \ge n + i$

In this case θ has a symbol of the form

$$\begin{pmatrix} u_1 & \dots & u_i & ua_1 & \dots & ua_n & \dots & u_{j+1} & \dots & u_{j+n} & \dots & u_m \\ 1 & \dots & 1 & 1 & \dots & 1 & \dots & 1 & \dots & 1 & \dots & 1 \\ v_1 & \dots & v_i & v_{i+1} & \dots & v_{i+n} & \dots & va_1 & \dots & va_n & \dots & v_m \end{pmatrix}.$$

Choose an ordered basis

$$w_1 < \dots < w_i < w < w_{i+n+1} < \dots < w_j < w' < w_{j+n+1} < \dots < w_m.$$

Then, if θ_1 and θ_2 are the elements with symbols

$$\begin{pmatrix} u_1 & \dots & u_i & u & u_{i+n+1} & \dots & u_{j+1} & \dots & u_{j+n} & \dots & u_m \\ 1 & \dots & 1 & 1 & 1 & \dots & 1 & \dots & 1 & \dots & 1 \\ w_1 & \dots & w_i & w & w_{i+n+1} & \dots & w'a_1 & \dots & w'a_n & \dots & w_m \end{pmatrix}$$

and

$$\begin{pmatrix} w_1 & \dots & w_i & wa_1 & \dots & wa_n & \dots & w_j & w' & w_{j+n+1} & \dots & w_m \\ 1 & \dots & 1 & 1 & \dots & 1 & \dots & 1 & 1 & 1 & \dots & 1 \\ v_1 & \dots & v_i & v_{i+1} & \dots & v_{i+n} & \dots & v_j & v & v_{j+n+1} & \dots & v_m \end{pmatrix},$$

respectively, we have $\theta_1, \theta_2 \in F_{n,r}$, both θ_1 and θ_2 have depth $d-1$, and

$$\theta = \theta_1\theta_2.$$

$j + 1 \le n + i$

In this case θ has a symbol of the form

$$\begin{pmatrix} u_1 & \dots & u_i & ua_1 & \dots\dots\dots ua_n \dots & u_{j+n} & u_{j+n+1} & \dots & u_m \\ 1 & \dots & 1 & 1 & \dots\dots\dots\dots\dots & 1 & 1 & \dots & 1 \\ v_1 & \dots & v_i & v_{i+1} & \dots va_1 \dots\dots\dots & va_n & v_{j+n+1} & \dots & v_m \end{pmatrix}.$$

We have that $m = d(n-1) + r \geq 4n - 4 + r$. First we suppose that $r \geq 2$ or $d \geq 5$, so that $m \geq 4n - 2$. Now, $j + 1 \leq n + i$, so if $j + n \geq m - n$ then $2n + i - 1 > m - n \geq 3n - 2$ and $i > n - 1$. So either $j + n \leq m - n$ or $i \geq n$. i.e. There are either at least n columns to the right of va_n or n columns to the left of ua_1. We will assume that $j + n \leq m - n$, the proof in the case $i \geq n$ is similar.

We can choose ordered bases

$$w_1 < \ldots < w_i < w < w_{i+n+1} < \ldots < w_{m-n} < w'$$

and

$$y_1 < \ldots < y_j < y < y_{j+n+1} < \ldots < y_{m-n} < y'.$$

Let θ_1, θ_2, and θ_3 be the elements with symbols

$$
\begin{pmatrix}
u_1 & \cdots & u_i & u & u_{i+n+1} & \cdots & u_{m-n} & u_{m-n+1} & \cdots & u_m \\
1 & \cdots & 1 & 1 & 1 & \cdots & 1 & 1 & \cdots & 1 \\
w_1 & \cdots & w_i & w & w_{i+n+1} & \cdots & w_{m-n} & w'a_1 & \cdots & w'a_m
\end{pmatrix}
$$

$$
\begin{pmatrix}
w_1 & \cdots & w_i & wa_1 & \cdots & \cdots & wa_n & \cdots & w_{j+n} & \cdots & w_{m-n} & w' \\
1 & \cdots & 1 & 1 & \cdots & \cdots & \cdots & \cdots & 1 & \cdots & 1 & 1 \\
y_1 & \cdots & y_i & y_{i+1} & \cdots & ya_1 & \cdots & \cdots & ya_n & \cdots & y_{m-n} & y'
\end{pmatrix}
$$

$$
\begin{pmatrix}
y_1 & \cdots & y_j & y & y_{j+n+1} & \cdots & y_{m-n} & y'a_1 & \cdots & y'a_n \\
1 & \cdots & 1 & 1 & 1 & \cdots & 1 & 1 & \cdots & 1 \\
v_1 & \cdots & v_j & v & v_{j+n+1} & \cdots & v_{m-n} & v_{m-n+1} & \cdots & v_m
\end{pmatrix},
$$

respectively. Then $\theta_1, \theta_2, \theta_3 \in F_{n,r}$, θ_1, θ_2, θ_3 have depth $d - 1$, and

$$\theta = \theta_1 \theta_2 \theta_3.$$

Finally suppose that $r = 1$ and $d = 4$. The proof goes through as above except in the case where $j + 1 = n + i$ and $i = n - 1$. In this case $m = 4n - 3$ and $j + n = m - (n - 1)$, so there are exactly $n - 1$ columns to the right of va_n and $n - 1$ to the left of ua_1. We choose ordered bases

$$w_1 < \ldots < w_{n-1} < w < w_{2n} < \ldots < w_{3n-3} < w',$$
$$y < y_{n+1} < y_{n+2} < \ldots < y_{3n-3} < y',$$

and

$$z < z_{n+1} < \ldots < z_{2n-2} < z' < z_{3n-1} < \ldots < z_{4n-3},$$

and let $\theta_1, \theta_2, \theta_3, \theta_4 \in F_{n,1}$ be the elements with symbols

$$
\begin{pmatrix}
u_1 & \cdots & u_{n-1} & u & u_{2n} & \cdots & u_{3n-2} & u_{3n-1} & \cdots & u_{4n-3} \\
1 & \cdots & 1 & 1 & 1 & \cdots & 1 & 1 & \cdots & 1 \\
w_1 & \cdots & w_{n-1} & w & w_{2n} & \cdots & w_{3n-2} & w'a_1 & \cdots & w'a_m
\end{pmatrix}
$$

$$\begin{pmatrix} w_1 & \cdots & w_{n-1} & wa_1 & wa_2 & \cdots & wa_n & w_{2n} & \cdots & w_{3n-3} & w' \\ 1 & \cdots & 1 & 1 & 1 & \cdots & 1 & 1 & \cdots & 1 & 1 \\ ya_1 & \cdots & ya_{n-1} & ya_n & y_{n+1} & \cdots & y_{2n-1} & y_{2n} & \cdots & y_{3n-3} & y' \end{pmatrix}$$

$$\begin{pmatrix} y & y_{n+1} & \cdots & y_{2n-1} & \cdots & y_{3n-3} & y'a_1 & y'a_2 & \cdots & y'a_n \\ 1 & 1 & \cdots & 1 & \cdots & 1 & 1 & 1 & \cdots & 1 \\ z & z_{n+1} & \cdots & z'a_1 & \cdots & z'a_{n-1} & z'a_n & z_{3n-1} & \cdots & z_{4n-m} \end{pmatrix}$$

$$\begin{pmatrix} za_1 & \cdots & za_{n-1} & za_n & z_{n+1} & \cdots & z_{2n-2} & z' & z_{3n-1} & \cdots & z_{4n-3} \\ 1 & \cdots & 1 & 1 & 1 & \cdots & 1 & 1 & 1 & \cdots & 1 \\ v_1 & \cdots & v_{n-1} & v_n & v_{n+1} & \cdots & v_{2n-2} & v & v_{3n-1} & \cdots & v_{4n-3} \end{pmatrix},$$

respectively. Then $\theta_1, \theta_2, \theta_3,$ and θ_4 have depth 3 and

$$\theta = \theta_1 \theta_2 \theta_3 \theta_4.$$

This proves the result. □

A *symbol for the relation* $\theta = \theta_1 \theta_2 ... \theta_t$ is an object of the form

$$\begin{pmatrix} y_{1,1} & y_{1,2} & \cdots & y_{1,m} \\ 1 & 1 & \cdots & 1 \\ y_{2,1} & y_{2,2} & \cdots & y_{2,m} \\ \vdots & \vdots & & \vdots \\ y_{t,1} & y_{t,2} & \cdots & y_{t,m} \\ 1 & 1 & \cdots & 1 \\ y_{t+1,1} & y_{t+1,2} & \cdots & y_{t+1,m} \end{pmatrix}$$

where

$$\begin{pmatrix} y_{i,1} & y_{i,2} & \cdots & y_{i,m} \\ 1 & 1 & \cdots & 1 \\ y_{i+1,1} & y_{i+1,2} & \cdots & y_{i+1,m} \end{pmatrix}$$

is a symbol for θ_i, $1 \leq i \leq t$. If $m = d(n-1) + r$ we say that the relation has depth d. So, if the relation has depth d so do θ and θ_i, $1 \leq i \leq t$. Notice, θ has symbol

$$\begin{pmatrix} y_{1,1} & y_{1,2} & \cdots & y_{1,m} \\ 1 & 1 & \cdots & 1 \\ y_{t+1,1} & y_{t+1,2} & \cdots & y_{t+1,m} \end{pmatrix}.$$

The next result can proved for the groups $T_{n,r}$, and $F_{n,r}$ by a similar proof to the proof given for $G_{n,r}$ in [5]. (We know [5] that any relation of depth $d \geq 7$ between elements of $G_{n,r}$ is a consequence of relations of depth less than d. It is just a matter of checking that these relations are relations between elements of $T_{n,r}$ or $F_{n,r}$ where appropriate, and this is straightforward as long as we take $d \geq 9$.)

LEMMA 14. *All the relations in each of $G_{n,r}$, $T_{n,r}$, and $F_{n,r}$ are consequences of their relations of depth at most 8.*

Theorem 15 does not follow immediately from Lemma 14 because there are infinitely many relations of depth eight. However, any relation of depth eight has a symbol whose rows have length at most $8(n-1) + r$. There are only finitely many distinct bases of size at most $8(n-1) + r$, so symbols for very long relations will contain repeated rows, and hence will be products of two shorter relations. Thus we may take those relations of depth eight whose length is less than some finite bound as generating relations for the groups, and we get:

THEOREM 15. *The groups $G_{n,r}$, $T_{n,r}$, $S_{n,r}$, and $F_{n,r}$ are finitely presented.*

The next theorem is proved in [2].

THEOREM 16. *$S'_{n,r}$ is finitely presented.*

Note, in general neither $S_{n,r}$ or $S'_{n,r}$ is the derived subgroup $T'_{n,r}$, although, if $n-1$ and r are co-prime $S_{n,r} = T_{n,r}$. In fact, $S'_{n,r} = T''_{n,r}$.

Next we consider some differences between the groups at which we have been looking. Theorem 18, which is about elements of finite order, can be contrasted with Theorem 7 to highlight some of the differences between the $G_{n,r}$, $T_{n,r}$, and $F_{n,r}$. The two results after that give some instances in which various different groups are isomorphic.

LEMMA 17. *An element of $\theta \in G_{n,r}$ has finite order if and only if it has a symbol whose top and bottom rows are permutatuions of each other.*

PROOF: One way around this result is obvious, so we will assume that $\theta^{t-1} = 1$ and show that θ has a symbol whose top and bottom rows are permutatuions of each other.

There are symbols

$$\begin{pmatrix} y_{i,1} & y_{i,2} & \cdots & y_{i,m} \\ 1 & 1 & \cdots & 1 \\ y_{i+1,1} & y_{i+1,2} & \cdots & y_{i+1,m} \end{pmatrix}$$

$1 \le i \le t-1$, for θ, such that

$$\begin{pmatrix} y_{1,1} & y_{1,2} & \cdots & y_{1,m} \\ 1 & 1 & \cdots & 1 \\ y_{2,1} & y_{2,2} & \cdots & y_{2,m} \\ \vdots & \vdots & & \vdots \\ y_{t,1} & y_{t,2} & \cdots & y_{t,m} \end{pmatrix}$$

is a symbol for the relation $\theta^{t-1} = 1$. So, in particular, $y_{1,j} = y_{t,j}$, $1 \le j \le m$.

Let

$$U_i = \{ y_{i,j} w \mid w \in W, \ 1 \le j \le m \},$$

so that $U_i \theta = U_{i+1}$ and $U_t = U_1$. Let

$$V = \bigcap_{1 \le i \le t} U_i,$$

then $V\theta = V$. Let $Y = \{y_1, \ldots, y_s\}$ be the basis of V, so that

$$V = \{ y_i w \mid w \in W, \ 1 \le i \le s \}.$$

Then $V\theta = V$, so $Y\theta = Y$ and θ has a symbol of the form

$$\begin{pmatrix} y_1 & \cdots & y_s \\ 1 & \cdots & 1 \\ y_{i_1} & \cdots & y_{i_s} \end{pmatrix},$$

as required.   □

THEOREM 18.

(i) $F_{n,r}$ has no non-trivial elements of finite order.

(ii) $T_{n,r}$ has elements of order t if and only if t divides some integer of the form $d(n-1)+r$, where $d \geq 0$. In particular, if $hcf(t, n-1)$ does not divide r, then $T_{n,r}$ has no elements of order t.

(iii) $S_{n,r}$ and $S'_{n,r}$ have elements of order t if and only if te divides some integer of the form $d(n-1)+r$, where $d \geq 0$.

The conditions in (ii) and (iii) are different, for example $T_{5,2}$ has elements of order 2 but the torsion elements of $S_{5,2}$ all have odd order.

PROOF: If $\theta \in T_{n,r}$ is an element of finite order then, by the above lemma, there exists a symbol

$$\Gamma \;=\; \begin{pmatrix} y_1 & \cdots & y_m \\ 1 & \cdots & 1 \\ y_{\pi(1)} & \cdots & y_{\pi(m)} \end{pmatrix}$$

for θ, where π is a permutation of $\{1, ..., m\}$. We can, and will, assume that $y_1 \leq ... \leq y_m$. Then, since $\theta \in T_{n,r}$, there exists some j such that

$$\Gamma \;=\; \begin{pmatrix} y_1 & \cdots & y_m \\ 1 & \cdots & 1 \\ y_{j+1} & \cdots & y_j \end{pmatrix}.$$

If $\theta \in F_{n,r}$ then $j = m$ and hence $\theta = 1$, proving (i).

The order of θ is the smallest integer, $t > 0$, such that m divides tj. For any integer t, if m divides tj but t does not divide m, then there is a smaller integer $t' > 0$, such that m divides $t'j$. So, if t is the order of some element $\theta \in T_{n,r}$, we must have that t divides m, for some $m = d(n-1) + r$. Conversely, if t divides $d(n-1)+r$, choose a basis $y_1 \leq ... \leq y_m$, where $m = d(n-1)+r$, and choose j so that $m = tj$. Then the element $\theta \in T_{n,r}$ which has symbol

$$\begin{pmatrix} y_1 & \cdots & y_m \\ 1 & \cdots & 1 \\ y_{j+1} & \cdots & y_j \end{pmatrix}$$

has order t. This proves (ii).

If $\theta \in S_{n,r}$ then θ has a symbol of the form

$$\Gamma \;=\; \begin{pmatrix} y_1 & \cdots & y_m \\ 1 & \cdots & 1 \\ y_{ej+1} & \cdots & y_{ej} \end{pmatrix}.$$

So exactly the same argument gives the result for elements of $S_{n,r}$. Finally, we know that the quotient group $S_{n,r}/S'_{n,r}$ is a free Abelian group, see Theorem 24, so all the elements of finite order in $S_{n,r}$ lie in $S'_{n,r}$.   □

Suppose that $X_r = \{x_1, ..., x_r\}$ and $X_{n-1+r} = \{x_1, ..., x_{n-1+r}\}$. We can define a map $\alpha : X_{n-1+r}W \rightarrow X_r W$ by

$$(x_i w)\alpha = \begin{cases} x_i w, & \text{for } 1 \le i \le r-1, \\ x_r a_j w, & \text{if } i = r+j-1, \ 1 \le j \le n. \end{cases}$$

Then α is an order-preserving bijection from $X_{n-1+r}W$ to $X_r W \setminus \{x_r\}$. If $\theta \in G_{n,n-1+r}$ has symbol

$$\begin{pmatrix} u_1 & \cdots & u_s \\ 1 & \cdots & 1 \\ v_1 & \cdots & v_s \end{pmatrix},$$

define $\theta_\alpha \in G_{n,r}$ to be the element with symbol

$$\begin{pmatrix} u_1\alpha & \cdots & u_s\alpha \\ 1 & \cdots & 1 \\ v_1\alpha & \cdots & v_s\alpha \end{pmatrix}.$$

It can be checked that θ_α is independent of the choice of symbol for θ, and that the map $\theta \rightarrow \theta_\alpha$ is an isomorphism which respects the order-preserving properties of θ. Thus we get:

THEOREM 19. If $r = d(n-1) + s$, for some integer d, then $G_{n,r} \cong G_{n,s}$, $T_{n,r} \cong T_{n,s}$, $S_{n,r} \cong S_{n,s}$, and $F_{n,r} \cong F_{n,s}$.

For the groups $F_{n,r}$ there is, see [2], a much stronger isomorphism result:

THEOREM 20. $F_{n,r} \cong F_{n,s}$, for all r and s, and $F_{n,r} \not\cong F_{m,s}$, if $n \ne m$.

We now describe the derived subgroups $F'_{n,r}$, $S'_{n,r}$, and $G'_{n,r}$ as kernels of specific homomorphisms from $F_{n,r}$, $S_{n,r}$, and $G_{n,r}$ into Abelian groups. Using these descriptions we show that $F'_{n,r}$, and $S'_{n,r}$ are simple.

Let $\mathbf{Z}^{n+1}$ be the free Abelian group of rank $n+1$, thought of as the set of vectors under addition generated by the elements $\varepsilon_i = (0 \ ... \ 0 \ 1 \ 0 \ ... \ 0)$, $1 \le i \le (n+1)$.

For any basis, Y, which is an expansion of χ_R we define $v_Y \in \mathbb{Z}^{n+1}$ inductively as follows:

(i) $v_{\chi_R} = 0 = (0\ 0\ ...\ 0)$.

(ii) If $Y = \{z_1 a_1, ..., z_1 a_n, z_2, ..., z_s\}$, where $z_1 \leq ... \leq z_s$, then $v_Y = v_Z + \varepsilon_1$.

(iii) If $Y = \{z_1, ..., z_{i-1}, z_i a_1, ..., z_i a_n, z_{i+1}, ..., z_s\}$, where $z_1 \leq ... \leq z_s$ and $2 \leq i \leq s - 1$, then $v_Y = v_Z + \varepsilon_t$, where $i \equiv t$ modulo $n - 1$ and $2 \leq t \leq n$.

(iv) If $Y = \{z_1, ..., z_{s-1}, z_s a_1, ..., z_s a_n\}$, where $z_1 \leq ... \leq z_s$, then $v_Y = v_Z + \varepsilon_{n+1}$.

Any expansion of a basis only changes the positions of the elements in the order modulo $n - 1$, so it is straightforward to check that the definition of v_Y is independent of the choice of simple contraction, Z, of Y taken.

We use the v_Y to define a homomorphism $f : F_{n,r} \to \mathbb{Z}^{n+1}$.

Suppose that $\theta \in F_{n,r}$ has a symbol,

$$\begin{pmatrix} u_1 & u_2 & ... & u_s \\ 1 & 1 & ... & 1 \\ v_1 & v_2 & ... & v_s \end{pmatrix},$$

whose top and bottom rows, U and V respectively, are expansions of χ_R. Define $\theta f = v_U - v_V$. Any symbol for θ whose rows are expansions of χ_R is an expansion of the unique shortest symbol

$$\begin{pmatrix} y_1 & y_2 & ... & y_m \\ 1 & 1 & ... & 1 \\ z_1 & z_2 & ... & z_m \end{pmatrix},$$

whose rows are expansions of χ_R. Thus we have that $v_U = v_Y + \varepsilon$ and $v_V = v_Z + \varepsilon$, for some $\varepsilon \in \mathbb{Z}^{n+1}$. So θf is independent of the choice of symbol used to define it. Since any two elements $\theta, \varphi \in F_{n,r}$ have symbols which can be combined, it is clear that f is a homomorphism.

If $v_Y = m_1 \varepsilon_1 + ... + m_{n+1} \varepsilon_{n+1}$, then Y is a basis of size $d(n-1) + r$, where $d = m_1 + ... + m_{n+1}$. If θ has a symbol with top and bottom rows Y and Z these bases must be of the same size. So we see that

$$m_1 \varepsilon_1 + ... + m_{n+1} \varepsilon_{n+1} \in \operatorname{Im} f \quad \text{only if} \quad m_1 + ... + m_{n+1} = 0.$$

It is easy to see that $\varepsilon_i - \varepsilon_{i+1}$ lies in $\operatorname{Im} f$, for $1 \leq i \leq n$, so $\operatorname{Im} f$ is the free Abelian subgroup of $\mathbb{Z}^{n+1}$ of rank n, which is freely generated by the elements $\varepsilon_i - \varepsilon_{i+1}$.

We write $F_{n,r}^+ = \ker f$.

Define $M_{n,r}$ to be the subgroup of $F_{n,r}$ consisting of all those elements that fix some $x_1 a_1^i$ and some $x_r a_n^j$. So $M_{n,r}$ is the group of all elements that have symbols of the form

$$\begin{pmatrix} y_1 & v_2 & \cdots & v_s & y_{s+1} \\ 1 & 1 & \cdots & 1 & 1 \\ y_1 & q_2 & \cdots & q_s & y_{s+1} \end{pmatrix}.$$

$M_{n,r}$ is the group called $F_{n,r}^0$ by Brown [2], who proves the following result:

LEMMA 21. $F_{n,r}^+ = M_{n,r}'$, the commutator subgroup of $M_{n,r}$.

We have seen that $\dfrac{F_{n,r}}{F_{n,r}^+} \cong \mathbf{Z}^n$, so $F_{n,r}' \subseteq F_{n,r}^+$. Hence $F_{n,r}' = M_{n,r}'$. We have introduced $M_{n,r}$ because it turns out to be easier to show that $M_{n,r}'$ is simple than to prove the result for $F_{n,r}^+$ directly.

We let λ_i denote the element of $F_{n,r}$ that has symbol

$$\begin{pmatrix} \xi_1 & \cdots & \xi_{i-1} & \xi_i a_1 & \xi_i a_2 & \cdots & \xi_i a_n & \xi_{i+1} & \xi_{i+2} & \cdots & \xi_R \\ 1 & \cdots & 1 & 1 & 1 & \cdots & 1 & 1 & 1 & \cdots & 1 \\ \xi_1 & \cdots & \xi_{i-1} & \xi_i & \xi_{i+1} a_1 & \cdots & \xi_{i+1} a_{n-1} & \xi_{i+1} a_n & \xi_{i+2} & \cdots & \xi_R \end{pmatrix},$$

$1 \le i \le R-1$, and λ_R denote the element of $T_{n,r}$ that has symbol

$$\begin{pmatrix} \xi_1 & \xi_2 & \cdots & \xi_{R-1} & \xi_R a_1 & \xi_R a_2 & \cdots & \xi_R a_n \\ 1 & 1 & \cdots & 1 & 1 & 1 & \cdots & 1 \\ \xi_1 a_n & \xi_2 & \cdots & \xi_{R-1} & \xi_R & \xi_1 a_1 & \cdots & \xi_1 a_{n-1} \end{pmatrix}.$$

It is obvious that $\lambda_i \mu = \mu \lambda_{i+1}$, $1 \le i \le R-1$, and that $\lambda_R \mu = \mu \lambda_1$. We also have, see Brown [2], that $\lambda_2 \lambda_3 \ldots \lambda_{e+1} \in S_{n,r}'$, $\lambda_2 \lambda_3 \ldots \lambda_n \in F_{n,r}'$, and that

$$\mu^{n-1} = \lambda_{R-n+1} \lambda_{R-n+2} \ldots \lambda_{R-1} \lambda_R \lambda_1 \lambda_2 \ldots \lambda_{R-1}.$$

Conjugating by μ^{R-n} we see that $\lambda_{R-n+2} \lambda_{R-n+3} \ldots \lambda_R \in S_{n,r}'$. Conjugating by suitable powers of μ and then taking products we see that

$\lambda_2\lambda_3 \ldots \lambda_R\lambda_1 \in S'_{n,r}$. Thus $\lambda_{R-n+1}^{-1}\mu^{n-1}\lambda_{R-n+1}$, and hence μ^{n-1}, lie in $S'_{n,r}$.

In fact, $F_{n,r}$ is generated by the elements $\lambda_1, \ldots, \lambda_{R-1}$, see [2], so $S_{n,r}$ is generated by the elements $\lambda_1, \ldots, \lambda_R$.

We also note here that since $e = hcf(n-1, R)$ and $\mu^R = 1$, μ^e is a power of μ^{n-1}. In particular, $\mu^e \in S'_{n,r}$.

For any element θ in $G_{n,r}$ we say that θ has *support* in V if $u \notin V$ implies that $u\theta = u$.

Given any non-trivial element, θ of $G_{n,r}$, we can find elements u and v such that $u\theta = v$ and u and v are independent, (i.e. $uW \cap vW = \emptyset$). For, we can choose a basis $\{u_1, ..., u_s\}$ in the domain of θ, and an integer i, such that $u_i\theta = v_i \neq u_i$. If u_i and v_i are independent then we are done. If not, one of u_i and v_i is a proper initial segment of the other. If u_i is a proper initial segment of v_i then, since $\{v_1, ..., v_s\}$ is a basis, there is some $j \neq i$ such that u_i is a proper initial segment of v_j. Since u_i and u_j are part of a basis they must be independent. Thus we can take $u = u_j$ and $v = v_j$.

Suppose that $\varphi \in G_{n,r}$ has support in uW, and that $u\theta = v$ where u and v are independent. Let $\psi = [\theta, \varphi] = \theta^{-1}\varphi^{-1}\theta\varphi$. Since ψ is maximal, and since $uW\theta^{-1} = vW$, it is easy to check that $uw\psi = uw\varphi$ and $vz\psi - vz(\theta^{-1}\varphi\theta)$, whenever the righthand sides are defined, and that ψ is trivial everywhere else.

We now give the simplicity results proved by Brown [2].

LEMMA 22. $F'_{n,r}$ *is a simple group.*

PROOF: Let N be a non-trivial subgroup of $F_{n,r}$ which is normalized by $F'_{n,r}$. We show that N contains $F'_{n,r}$.

Pick a non-trivial element $\theta_1 \in N$, and choose $u_0 \in XW$ such that u_0 and $v_0 = u_0\theta_1$ are independent. Taking u_0w and v_0w if necessary, we may suppose that $v_0\theta_1$ is defined. So there is a basis in the domain of θ_1 which contains them both. Let $w_0 = v_0\theta_1$, then w_0 and v_0 are independent since they lie in a basis. Since θ_1 is order preserving, either $u_0 < v_0 < w_0$ or $u_0 > v_0 > w_0$, and so u_0 and w_0 must also be independent. Let $\theta_2 = \theta_1^2$, so $u_0\theta_2 = w_0$.

First we show that if $\varphi_1, \varphi_2 \in F_{n,r}$ have support in u_0W, then $[\varphi_1, \varphi_2]$ lies in N.

From the remarks make above this lemma we see that $\rho_i = [\theta_i, \varphi_i]$ is equal to φ_i and $\theta_i^{-1}\varphi_i\theta_i$ on their (disjoint) supports and is trivial everywhere else. Notice also that $\rho_i \in N$. Since the supports of $\theta_1^{-1}\varphi_1\theta_1$ and $\theta_2^{-1}\varphi_2\theta_2$ lie in v_0W and w_0W respectively, they are disjoint. Hence $[\rho_1, \rho_2]$ is equal to $[\varphi_1, \varphi_2]$ on the support of the latter, and is trivial everywhere else. i.e. $[\varphi_1, \varphi_2] = [\rho_1, \rho_2] \in N$.

(Notice, u_0 cannot be the first or last element of any ordered basis.)

Now, given any two elements $\varphi, \psi \in M_{n,r}$, we can find symbols of the form

$$\begin{pmatrix} y_1 & v_2 & \cdots & v_s & y_{s+1} \\ 1 & 1 & \cdots & 1 & 1 \\ y_1 & q_2 & \cdots & q_s & y_{s+1} \end{pmatrix},$$

$$\begin{pmatrix} y_1 & v_2 & \cdots & v_s & y_{s+1} \\ 1 & 1 & \cdots & 1 & 1 \\ y_1 & z_2 & \cdots & z_s & y_{s+1} \end{pmatrix},$$

respectively, for these elements. Expanding y_1 and y_{s+1} if necessary, we may assume that there is an ordered basis of the form

$$y_1 < \cdots < y_{i-1} < u_0 < y_{i+1} < \cdots < y_{s+1}.$$

Let

$$Z = \{y_1a_1, ..., y_1a_n, v_2, ..., v_s, y_{s+1}a_1, ..., y_{s+1}a_n\},$$

and $Y = \{y_1a_1, ..., y_1a_n, y_2, ..., y_s, y_{s+1}a_1, ..., y_{s+1}a_n\}.$

Expand Z to $Z_0 = (Z \backslash \{y_1a_n\}) \cup \{y_1a_nu_1, ..., y_1a_nu_\alpha\}$ and then Y to $Y_1 = (Y \backslash \{u_0\}) \cup \{u_0w_1, ..., u_0w_t\}$, where $\alpha \geq i-1$ and $i-1+t \geq \alpha+s$. Then expand Z_0 to

$$Z_1 = (Z_1 \backslash \{y_{s+1}a_1\}) \cup \{y_{s+1}a_1z_1, ..., y_{s+1}a_1z_\beta\},$$

where $\beta = t - \alpha$. Let τ_1 be the element of $F_{n,r}$ which carries the ordered basis Z_1 to the ordered basis Y_1. So τ_1 has a symbol of the form

$$\begin{pmatrix} y_1a_1 & \cdots & \cdots y_1a_nu_\alpha & v_2 & \cdots & v_s & y_{s+1}a_1z_1 \cdots & \cdots & y_{s+1}a_n \\ 1 & \cdots & \cdots\cdots\cdots\cdots & \cdots & \cdots & \cdots & \cdots\cdots\cdots\cdots & \cdots & 1 \\ y_1a_1 & \cdots & u_0w_1\cdots\cdots & \cdots & \cdots & \cdots & \cdots\cdots\cdots u_0w_t & \cdots & y_{s+1}a_n \end{pmatrix}.$$

Thus τ_1 carries the supports of φ and ψ into u_0W.

Re-write the symbol above for τ_1 in the form

$$\begin{pmatrix} y_1a_1 & f_2 & \cdots & f_\gamma & y_{s+1}a_n \\ 1 & 1 & \cdots & 1 & 1 \\ y_1a_1 & p_2 & \cdots & p_\gamma & y_{s+1}a_n \end{pmatrix}.$$

Expanding $y_1 a_1 a_n$ on the top row and $y_{s+1} a_n a_1$ on the bottom row, we can see that there exists an element $\tau_2^{-1} \in F_{n,r}$ that has a symbol of the form

$$\begin{pmatrix} y_1 a_1 a_1 & \cdots & \cdots & \cdots & \cdots & f_2 & \cdots & f_\gamma & \cdots & y_{s+1} a_n \\ 1 & \cdots & \cdots & \cdots & \cdots & \cdots & \cdots & \cdots & \cdots & 1 \\ y_1 a_1 & p_2 & \cdots & p_\gamma & \cdots & \cdots & \cdots & \cdots & \cdots & y_{s+1} a_n a_n \end{pmatrix}.$$

Then $\tau = [\tau_2, \tau_1]$ lies in $F'_{n,r}$ and agrees with τ_1 on $\{f_2, ..., f_\gamma\}$. So $\tau^{-1} \varphi \tau$ and $\tau^{-1} \psi \tau$ have supports in $u_0 W$, and hence by the first part,

$$[\varphi, \psi] = \tau [\tau^{-1} \varphi \tau, \tau^{-1} \psi \tau] \tau^{-1} \in \tau^{-1} N \tau = N.$$

Thus $F'_{n,r} = M'_{n,r} \subseteq N$, as required.

THEOREM 23. $S'_{n,r}$ is a simple group.

PROOF: Let N be a non-trivial subgroup of $S_{n,r}$ which is normalized by $S'_{n,r}$. We show that N contains $S'_{n,r}$.

Since μ^e is a power of μ^{n-1}, and since $\lambda_i \in F_{n,r}$, for $i \neq R$, $S_{n,r}$ is generated by $F_{n,r}$ and λ_R. Any commutator subgroup is generated by conjugates of commutators of generators, so it is enough to show that $[\delta_1, \delta_2]^\psi \in N$, for all $\delta_1, \delta_2 \in F_{n,r} \cup \{\lambda_R\}$, and for all $\psi \in S_{n,r}$.

Let ψ be any element of $S_{n,r}$, and let θ be a non-trivial element of N. First we show that $\psi N \psi^{-1}$ contains a non-trivial element of $F_{n,r}$.

Suppose that $\psi \theta \psi^{-1}$ has symbol

$$\begin{pmatrix} u_1 & u_2 & \cdots & u_s \\ 1 & 1 & \cdots & 1 \\ v_{j+1} & v_{j+2} & \cdots & v_j \end{pmatrix}.$$

If $j = 0$ then $\psi \theta \psi^{-1} \in F_{n,r}$. So suppose that $j \geq 1$. Then we can expand the symbol if necessary to insure that if $v_1 = \xi_1 a_1^m$ then $u_1 \neq \xi_1 a_1^{2m}$. Now $\theta = \delta \mu^i \alpha$, for some i which is congruent to j modulo $n - 1$. Let $\gamma = \delta \alpha$ and let $\tau = \alpha^{-1} \mu^i \alpha$. Then $\gamma \neq 1$, so γ has infinite order, and e divides i, so $\tau \in S'_{n,r}$. Then $\psi \tau \psi^{-1} \in S'_{n,r}$, so τ normalizes $\psi N \psi^{-1}$, and hence, for any m, $\gamma^m \tau^m = \gamma \tau \tau^{-1} \gamma \tau^2 \tau^{-2} \ldots \gamma \tau^m \in \psi N \psi^{-1}$. Since τ has finite order, for some m, $1 \neq \gamma^m \in \psi N \psi^{-1} \cap F_{n,r}$.

Then $\psi N \psi^{-1} \cap F_{n,r}$ is normalized by $F'_{n,r}$, and so, from the proof of Lemma 22, we have that $\psi^{-1} F'_{n,r} \psi \subseteq N$.

Since $2 \leq R - e \leq R$, it is easy to check that $[\lambda_{R-e}, \lambda_R] = 1$. Since $[\lambda_i, \lambda_R]^\psi = [\lambda_{i+e}, \lambda_e]^{\mu^{-e}\psi}$ we have that $[\lambda_R, \lambda_i]^{-\psi} = [\lambda_i, \lambda_R]^\psi \in N$, as long as $i \neq R - e$. Then, since $F_{n,r}$ is generated by the λ_i, $i \neq R$, we see that $S'_{n,r} \subseteq N$, as required. □

From the above result and Theorem 16 we see that $S'_{n,r}$ is a finitely presented (infinite) simple group. Note, from Theorem 18 and Theorem 7 we see that $S'_{n,r}$ is different from any $G^+_{m,s}$ if $n - 1$ does not divide r.

Using a similar construction to that given below Theorem 18, we can define a homomorphism $f : S_{n,r} \to Z^e$ where Z^e is the free Abelian group of rank e, generated by the elements $\varepsilon_i = (0 \ldots 0\, 1\, 0 \ldots 0)$, $1 \leq i \leq e$.

We define $u_{\chi_R} = 0 = (0\, 0 \ldots 0)$. Then, for any basis, Y, which is an expansion of χ_R we define $u_Y \in Z^e$ to be $u_Y = u_Z + \varepsilon_t$, where

$$Y = \{z_1, \ldots, z_{i-1}, z_i a_1, \ldots, z_i a_n, z_{i+1}, \ldots, z_s\},$$

$z_1 \leq \ldots \leq z_s$ and $i \equiv t$ modulo e. (So in this case expansions from the first or last elements are not treated any differently to any other expansion.) Since e divides $n - 1$ and R, the definition of u_Y is independent of the choice of simple contraction, Z, of Y taken.

We define $f : S_{n,r} \to Z^e$ by setting $\theta f = u_U - u_V$, where $\theta \in S_{n,r}$ has a symbol,

$$\begin{pmatrix} u_1 & u_2 & \ldots & u_s \\ 1 & 1 & \ldots & 1 \\ v_{j+1} & v_2 & \ldots & v_j \end{pmatrix},$$

whose top and bottom rows, U and V respectively, are expansions of χ_R. Since e divides j and expansions on first and last elements are not treated differently, it can be seen, as for the case $f : F_{n,r} \to Z^{n+1}$, that f is a well defined homomorphism, and that the image of f is a free Abelian group of rank $e - 1$.

It is shown in [2] that $S'_{n,r} = \ker f$.

For completeness we can give a similar description of $G^+_{n,r} = G'_{n,r}$.

Let $c = hcf(n - 1, 2)$. For any basis, Y, which is an expansion of χ_R, define $u_Y \in Z_c$ to be $u_Y = u_Z + t$, where

$$Y = \{z_1, \ldots, z_{i-1}, z_i a_1, \ldots, z_i a_n, z_{i+1}, \ldots, z_s\},$$

$z_1 \leq \ldots \leq z_s$ and $i \equiv t$ modulo c. We define $f : G_{n,r} \to Z_c$ by setting $\theta f = u_U - u_V$, where $\theta \in G_{n,r}$ has a symbol whose top and bottom rows, U

and V respectively, are expansions of χ_R. Then f is a well defined surjective whose kernel, $G_{n,r}^+$, is a simple group.

Thus we have the following result:

THEOREM 24. *Let* $e = hcf(r, n-1)$ *and* $c = hcf(2, n-1)$, *then*

$$\frac{F_{n,r}}{F'_{n,r}} \cong \mathbf{Z}^n, \quad \frac{S_{n,r}}{S'_{n,r}} \cong \mathbf{Z}^{e-1}, \quad \frac{T_{n,r}}{S_{n,r}} \cong \mathbf{Z}_e, \quad \text{and} \quad \frac{G_{n,r}}{G'_{n,r}} \cong \mathbf{Z}_c.$$

5. Subgroups Of $\mathcal{G}_{\backslash,\infty}$ That Contain $G_{n,1}^+$

For the rest of the paper we will consider finitely presented groups which contain some $G'_{n,1}$. In this section we will describe explict sets of defining relations for H-expansible groups.

The following theorem was proved for the case $n = 2$, $r = 1$ by Thompson [11]. The proof for general n when $r = 1$ is written out in [8], and it is easy to see that this proof can be adapted to prove the general result:

THEOREM 25.   *If* K *is any group such that* $G_{n,r}^+ \leq K \leq \mathcal{G}_{n,r}$ *then the derived subgroup,* K', *is simple.*

From now on we shall only consider the case $r = 1$. We shall adopt the convention, described above, of identifying $x_1 w$ with w, for all $w \in W$.
    For any subgroup, H, of $\mathcal{G}_{n,1}$, let

$$\mathcal{H} = \langle G_{n,1}, H \rangle.$$

For $h \in H$, let σ_h be the element of $\mathcal{G}_{n,1}$ which has H-symbol

$$\begin{pmatrix} a_1 & a_2 & \cdots & a_n \\ h & 1 & \cdots & 1 \\ a_1 & a_2 & \cdots & a_n \end{pmatrix}.$$

It can be shown [8] that

$$\mathcal{H} = \langle G_{n,1}, \sigma_h \mid h \in H \rangle.$$

Suppose that $\mathcal{H}$ is H-expansible. Let G_1 and H_1 be sets of generators for $G_{n,1}$ and H, respectively. Let $G^\sharp$ be the set of all elements, $\alpha \in G_{n,1}$, such that $a_1\alpha = a_1$. Let δ be the element of $G_{n,1}$ which has symbol

$$\begin{pmatrix} a_1 & a_2 & \cdots & a_{n-1} & a_n a_1 & a_n a_2 & \cdots & a_n a_n \\ 1 & 1 & \cdots & 1 & 1 & 1 & \cdots & 1 \\ a_n a_1 & a_2 & \cdots & a_{n-1} & a_1 & a_n a_2 & \cdots & a_n a_n \end{pmatrix}.$$

Finally, we say that a relation $\sigma_h = \tau\sigma_{h_1}\eta_2\sigma_{h_2}\cdots\eta_s\sigma_{h_s}\eta_s\cdots\eta_3\eta_2\varepsilon$ is a simple defining relation for σ_h if

(i) σ_h has an H-symbol $\begin{pmatrix} u_1 & u_2 & \cdots & u_s \\ h_1 & h_2 & \cdots & h_s \\ v_1 & v_2 & \cdots & v_s \end{pmatrix}$,

(ii) For some basis $\{a_1, w_2, \ldots, w_s\}$, τ, ε, and η_i have symbols

$$\begin{pmatrix} u_1 & u_2 & \cdots & u_s \\ 1 & 1 & \cdots & 1 \\ a_1 & w_2 & \cdots & w_s \end{pmatrix}, \quad \begin{pmatrix} a_1 & w_2 & \cdots & w_s \\ 1 & 1 & \cdots & 1 \\ v_1 & v_2 & \cdots & v_s \end{pmatrix},$$

and $\begin{pmatrix} a_1 & w_2 & \cdots & w_{i-1} & w_i & w_{i+1} & \cdots & w_s \\ 1 & 1 & \cdots & 1 & 1 & 1 & \cdots & 1 \\ w_i & w_2 & \cdots & w_{i-1} & a_1 & w_{i+1} & \cdots & w_s \end{pmatrix},$

respectively.

Taking G_1 and $\{\sigma_h \mid h \in H_1\}$ as generators, it can be shown [8] that, if $\mathcal{H}$ is H-expansible, then $A \cup B \cup C \cup D$ is an (infinite) set of defining relations for $\mathcal{H}$, where

(A) A set of defining relations for $G_{n,1} = \langle G_1 \rangle$, and a set for $\sigma_H = \langle H_1 \rangle$

(B) $\{\alpha\sigma_h = \sigma_h\alpha \mid h \in H, \ \alpha \in G^\sharp\}$

(C) $\{\delta\sigma_h\delta\sigma_k = \sigma_k\delta\sigma_h\delta \mid h, k \in H\}$.

(D) The set of all simple defining relations for σ_h, $h \in H$.

(The above relations are written in terms of general elements of $G_{n,1}$ and σ_H, not just elements from the chosen generating set. The elements are considered as 'shorthand' notation for words on the generators.)

It is not easy, in general, to find groups $\mathcal{H}$ which are H-expansible. It is not always even the case that H is H-expansible, some elements of H may have symbols which cannot be combined.

EXAMPLE: Let $r = 1$ and $n = 2$, let $a_1 = a$ and $a_2 = b$, and let t be the maximal inescapable isomorphism defined by the rules

$$a^{2i+1}b \rightarrow a^{2i}b$$
$$a^{2i}b \rightarrow a^{2i+1}b.$$

Let s be the maximal inescapable isomorphism defined by the rules

$$a^{2i+1}b \rightarrow a^{2i+2}b$$
$$a^{2i+2}b \rightarrow a^{2i+1}b$$
$$b \rightarrow b.$$

Then let $H = \langle t, s \rangle$. The elements t and s have H-symbols

$$\begin{pmatrix} aa & ab & b \\ t & 1 & 1 \\ aa & b & ab \end{pmatrix} \quad \text{and} \quad \begin{pmatrix} a & b \\ t & 1 \\ a & b \end{pmatrix},$$

respectively. Any expansions of these are are of the forms

$$\begin{pmatrix} a^{2i} & v_2 & \cdots & v_s \\ t & 1 & \cdots & 1 \\ a^{2i} & u_2 & \cdots & u_s \end{pmatrix} \quad \text{and} \quad \begin{pmatrix} a^{2i+1} & v_2 & \cdots & v_s \\ t & 1 & \cdots & 1 \\ a^{2i+1} & u_2 & \cdots & u_s \end{pmatrix},$$

which cannot be combined.

(In fact, ts cannot have an H-symbol of any form. Suppose that Γ is an H-symbol for ts, and that Γ contains a^{2i+2} in its top row. It must also then have an element of the form $a^{2i}bw$ in its top row. Since $(a^{2i}bw)ts = a^{2i+2}bw$, there must be an element of the form $a^{2i+1}bu$ in the bottom row. Thus there must be an element of the form $(a^{2i+1}bu)s^{-1}t^{-1} = a^{2i+3}bu$ in the top row, which is not possible since a basis cannot contain both a^{2i+2} and $a^{2i+3}bu$. A similar argument deals with the case a^{2i+1}.)

6.   Wreath Products

We now consider a class of groups, H, for which $\mathcal{H}$ is always H-expansible, and furthermore, for which if H is finitely presented then so is $\mathcal{H}$.

Let A and B be permutation groups on the sets Δ and Σ, respectively. Let A^Σ denote the set of maps from Σ into A. Then the restricted wreath product $AwrB = A^\Sigma B$ is defined as a permutation group on the set $\Delta \times \Sigma$ by the rule

$$(\delta, \gamma)fb = (\delta f(\gamma), \gamma b),$$

for all $f \in A^\Sigma$ and all $b \in B$.

There is a natural projection, $\pi : AwrB \to B$ given by $(fb)\pi = b$. If C is a permutation group on Λ, and if $\theta : A \to C$ is a homomorphism, then there is a natural homomorphism $\theta' : AwrB \to CwrB$ given by

$$(\delta, \gamma)(fb)\theta' = (\delta(f(\gamma))\theta, \gamma b).$$

Let F be a free group acting on itself by right multiplication, and let S_{n-1} denote the full symmetric group on the set $\{a_1, \ldots, a_{n-1}\}$, and write

$$F\,(\mathrm{wr}\ S_{n-1})^i = (..((F\mathrm{wr}S_{n-1})\,\mathrm{wr}S_{n-1})\,\mathrm{wr}...)\,\mathrm{wr}S_{n-1},$$

$$S_{n-1}\,(\mathrm{wr}\ S_{n-1})^i = (..((S_{n-1}\mathrm{wr}S_{n-1})\,\mathrm{wr}S_{n-1})\,\mathrm{wr}...)\,\mathrm{wr}S_{n-1}.$$

Suppose that we are given a homomorphism $\theta : F \to F\mathrm{wr}S_{n-1}$. Let $\theta_0 = \theta$, and inductively define the induced homomorphisms,

$$\theta_i = \theta'_{i-1} : F\,(\mathrm{wr}\ S_{n-1})^i \to F\,(\mathrm{wr}\ S_{n-1})^{i+1},$$

as above. Then we get a commuting ladder of homomorphisms

$$
\begin{array}{ccccccccc}
F & \xrightarrow{\theta} & F\mathrm{wr}S_{n-1} & \xrightarrow{\theta_1} & \cdots \longrightarrow & F(\mathrm{wr}S_{n-1})^i & \xrightarrow{\theta_i} & F(\mathrm{wr}S_{n-1})^{i+1} & \longrightarrow \cdots \\
& & \pi_1 \downarrow & & & \pi_i \downarrow & & \pi_{i+1} \downarrow & \\
& & S_{n-1} & \xleftarrow{\varphi_1} & \cdots \longleftarrow & S_{n-1}(\mathrm{wr}S_{n-1})^{i-1} & \xleftarrow{\varphi_i} & S_{n-1}(\mathrm{wr}S_{n-1})^i & \longleftarrow \cdots
\end{array}
$$

where π_i and φ_i are the natural projections. From the ladder, we can read off homomorphisms

$$\psi_i : F \to S_{n-1}(\mathrm{wr}S_{n-1})^{i-1}$$

which are such that $\ker\psi_{i+1} \subseteq \ker\psi_i$. We define

$$\Psi = \bigcap_i \ker\psi_i,$$

and consider the group F/Ψ, which is called *the group defined by θ.*

For $L \in F$, let ℓ be the maximal inescapable isomorphism defined by the rules

$$ua_n \rightarrow (u(L\psi_m))a_n$$

$$a_n \rightarrow a_n,$$

where u is any element of W, of length m, which does not contain any occurence of a_n. It is not hard to check that the map $\Theta : F \rightarrow \mathcal{G}_{n,1}$ given by $L\Theta = \ell$ is a homomorphism with kernel Ψ. We let H_θ denote the image of Θ, and call H_θ the subgroup of $\mathcal{G}_{n,1}$ defined by θ.

Because of the way symbols for elements of H_θ can be expanded, we call groups of the form H_θ *groups defined by permutation expansion rules.*

The groups H_θ are H_θ-expansible: For $L \in F$, we can think of $L\theta$ as an ordered n-tuple, $(L_1, ..., L_{n-1}, \pi)$, where $L_1 \in F$ and $\pi \in S_{n-1}$. Then ℓ has H_θ-symbol

$$\begin{pmatrix} a_1 & \cdots & a_{n-1} & a_n \\ \ell_1 & \cdots & \ell_{n-1} & 1 \\ a_1\pi & \cdots & a_{n-1}\pi & a_n \end{pmatrix}.$$

The next two results are proved in [8], and provide a strategy for embedding certain types of group into finitely presented simple groups. Theorem 27 shows that if H_θ is finitely presented then H'_θ is embeddable in a finitely presented simple group.

LEMMA 26.   *The groups $\langle \mathcal{G}_{n,1}, H_\theta \rangle$ are H_θ-expansible, and, if H_θ is finitely presented, we can choose a finite subset of the relations $A \cup B \cup C \cup D$ to define $\langle \mathcal{G}_{n,1}, H_\theta \rangle$.*

THEOREM 27.   *Let $\theta : F \rightarrow F\mathrm{wr}S_{n-1}$ be a homomorphism, and let H_θ be the subgroup of $\mathcal{G}_{n,1}$ defined by θ. If H_θ is finitely presented then there exists an integer m, and a homomorphism $\varphi : F \rightarrow F\mathrm{wr}S_{n-1}$ such that $H_\theta \cong H_\varphi$ and $\langle \mathcal{G}_{m,1}, H_\varphi \rangle'$ is a finitely presented simple group.*

In view of this, it would be interesting to know what types of finitely presented groups are of the form H_θ. We do know that they are all residually finite. In fact, given any non-trivial element $\ell \in H_\theta$ there exists a homomorphism $f : H_\theta \rightarrow S_{n-1}(\mathrm{wr}S_{n-1})^i$ such that $\ell f \neq 1$. All finitely generated 1-relator groups have solvable word problem, but there exist examples of such groups which are not residually finite. Thus we cannot hope to embed all finitely presented groups with solvable word problem into finitely presented simple groups using only groups defined by permutation expansion rules.

Together with example below Theorem 8, the next theorem shows that we do get some new finitely presented infinite simple groups by the above type of construction.

THEOREM 28.   For each integer n, there exists an integer m such that $\langle G_{m,1},\ \mathbf{Z}^n.\mathrm{GL}(n,\mathbf{Z})\rangle$ is a finitely presented simple group.

EXAMPLE: Let $F = \mathrm{free}\langle T, S\rangle$, let $n = 6$, let m be any integer and let $\pi_T = (a_1\ a_2)(a_3\ a_4)$, $\pi_S = 1$. Write $f\pi \in FwrS_5$ as an ordered 6-tuple, $(f(a_1), f(a_2), f(a_3), f(a_4), f(a_5), \pi)$.

Define $\theta : F \rightarrow FwrS_5$ by the rules

$$T \rightarrow (T, 1, T, 1, 1, \pi_T)$$
$$S \rightarrow (S, ST^{-m}, S, ST^{-m}, 1, 1).$$

Then $H_\theta = \langle t, s\rangle$, where t and s have symbols

$$\begin{pmatrix} a_1 & a_2 & a_3 & a_4 & a_5 & a_6 \\ t & 1 & t & 1 & 1 & 1 \\ a_2 & a_1 & a_4 & a_3 & a_5 & a_6 \end{pmatrix}, \qquad \begin{pmatrix} a_1 & a_2 & a_3 & a_4 & a_5 & a_6 \\ s & st^{-m} & s & st^{-m} & 1 & 1 \\ a_1 & a_2 & a_3 & a_4 & a_5 & a_6 \end{pmatrix},$$

respectively. Then, if $q = 2m + 1$, $s^{-1}tst^{-q}$ has symbol

$$\begin{pmatrix} a_1 & a_2 & a_3 & a_4 & a_5 & a_6 \\ s^{-1}tst^{-q} & 1 & s^{-1}tst^{-q} & 1 & 1 & 1 \\ a_1 & a_2 & a_3 & a_4 & a_5 & a_6 \end{pmatrix}.$$

So, using the maximality of elements of $\mathcal{G}_{n,1}$, we have that $s^{-1}tst^{-q} = 1$. It is not hard to see that t must have infinite order, and hence that

$$H_\theta \ = \ \langle t, s \mid s^{-1}tst^{-q}\rangle.$$

It is possible, [9], to find a homomorphism $\theta : F \rightarrow FwrS_{q+2}$, where q is even, such that $H_\theta = \langle t, s \mid s^{-1}tst^{-q}\rangle$. This allows the proof of the following result:

THEOREM 29.   *For any integer q, the additive group*

$$\left\{ \frac{c}{q^b} \ \Big| \ c, b \in \mathbb{Z} \right\}$$

can be embedded in a finitely presented simple group.

In fact, a stronger result is proved in [9]. Let q be any integer, and let Q be the ring of rationals whose denominators are powers of q. Then:

THEOREM 30.   *If A is a countable Abelian group whose torsion factor group is a subgroup of some countable direct sum of cyclic Q-modules, then A can be embedded in a finitely presented simple group.*

NOTE: I have tried to extend the above techniques for embedding Abelian groups to embed the group $\mathbb{Q}^+$ in a finitely presented simple group, but I have met with little success.

7.   The Conjugacy Problem

Suppose that L is a finitely presented subgroup of $G_{n,1}$, where $n = 2s+1$, and that θ_i are automorphisms of L, for $2 \leq i \leq s$. Form the split extension of L by $\langle \theta_2, \ldots \theta_s \rangle$ in the usual way, so that $\theta_i^{-1} \ell \theta_i$ is the image of ℓ under θ_i. It can be shown, see [10], that there is an embedding

$$f : L \rtimes \langle \theta_2, \ldots \theta_s \rangle \quad \rightarrow \quad \mathcal{G}_{n,1}$$

such that , if $H = (L \rtimes \langle \theta_2, \ldots \theta_s \rangle)f$ then $\langle G_{n,r} \ , \ H \rangle$ is H-expansible. (The group H is not a group defined by permutation expansion rules.)

It can also be shown, since L is finitely presented, that the defining relations $A \cup B \cup C \cup D$ for $\mathcal{H}$ described above are generated by a finite subset. Thus $\mathcal{H}$ is finitely presented.

It is fairly easy to see that for $\ell_1, \ell_2 \in L$, $\sigma_{\ell_1 f}$ and $\sigma_{\ell_2 f}$ are conjugate in σ_H if and only if they are conjugate in $\mathcal{H}$.

C.F. Miller has given a construction of a group of the form $L \rtimes \langle \theta_2, \ldots \theta_s \rangle$ which has unsolvable conjugacy problem. The group L is a free group of rank p and $s = q - p$, where p and q come from a presentation

$$\langle g_2, \ldots, g_p \ | \ R_1 = \ldots R_q = 1 \rangle$$

of a finitely presented torsion free group with unsolvable word problem, see [7]. Every free group of finite rank can be embedded in every $G_{n,r}$, see [10, Appendix]. So we can construct a finitely presented subgroup $\mathcal{H} \subseteq \mathcal{G}_{n,r}$ which inherits an unsolvable conjugacy problem from H.

The derived subgroup, $\mathcal{H}'$, is of finite index in $\mathcal{H}$ and inherits an unsolvable conjugacy problem from $\mathcal{H}$. Thus $\mathcal{H}'$ is a finitely presented simple group with an unsolvable conjugacy problem.

REFERENCES

1. W.W. Boone and G. Higman, *An algebraic characterization of the solvability of the word problem*, J. Austral. Math. Soc. **18** (1974), 41–53.

2. K. Brown, *Finiteness properties of groups*, J. Pure Appl. Algebra **44** (1987), 45–75.

3. G. Higman, B.H. Neumann, and H. Neumann, *Embedding theorems for groups*, J. London Math. Soc. **24** (1949), 247–254.

4. G. Higman, *Subgroups of finitely presented groups*, Proc. Royal Soc. London Ser. A **262** (1961), 455–475.

5. G. Higman, "Finitely Presented Infinite Simple Groups", Notes on Pure Mathematics 8, Australian National University, Canberra, 1974.

6. A.V. Kuznetsov, *Algorithms as operations in algebraic systems*, Izv. Akad. Nauk SSSR Ser. Mat. (1958).

7. C.F. Miller III, *Decision problems in algebraic classes of groups (a survey)*, "Word Problems", (edited by W.W. Boone, F.B. Cannonito, and R.C. Lyndon), North Holland, Amsterdam, 1973.

8. E.A. Scott, *A construction which can be used to produce finitely presented infinite simple groups*, Journal of Algebra **90** (1984), 294–322.

9. E.A. Scott, *The embedding of certain linear and abelian groups in finitely presented simple groups*, Journal of Algebra **90** (1984), 323–332.

10. E.A. Scott, *A finitely presented simple group with an unsolvable conjugacy problem*, Journal of Algebra **90** (1984), 333–353.

11. R.J. Thompson, *Embeddings into finitely generated groups which preserve the word problem*, "Word Problems II", (edited by S.I. Adian, W.W. Boone, and G. Higman), North-Holland, Amsterdam, 1980.

School Mathematical Sciences, The Faculties, Australian National University, Canberra, 2601, Australia

The Geometry of Finitely Presented Infinite Simple Groups

KENNETH S. BROWN

Abstract. Let $\mathcal{G}$ be the family of finitely presented infinite simple groups introduced by Higman, generalizing R.J. Thompson's group of dyadic homeomorphisms of the Cantor set. For each $G \in \mathcal{G}$ and each integer $n \geq 1$, an $(n-1)$-connected n-dimensional simplicial complex is constructed, on which G acts with finite stabilizers and with an n-simplex as fundamental domain. This yields homological and combinatorial information about G. As a by-product, one obtains a solution to a problem of Neumann and Neumann.

Introduction

In 1965 R.J. Thompson gave the first example of a finitely presented infinite simple group (cf. [7]). It can be described as a group of homeomorphisms of the Cantor set or, alternatively, as a certain algebraic automorphism group. Higman [6] later introduced an infinite family $\mathcal{G}$ of finitely presented infinite simple groups generalizing Thompson's example. It is shown in [4] that each $G \in \mathcal{G}$ has the homological finiteness property FP_∞, i.e., that $\mathbf{Z}$ admits a resolution by finitely generated free $\mathbf{Z}G$-modules. The proof is topological; it involves the construction of highly connected simplicial complexes X such that G acts on X with finite stabilizers and compact quotient. But the complexes X are complicated, and it is difficult to obtain any information about G from them other than the FP_∞ property. The purpose of the present note is to give a better construction. Recall that a space X is said to be $(n-1)$-*connected* if $\pi_i(X) = 0$ for $i < n$.

MAIN THEOREM. *For any $G \in \mathcal{G}$ and any $n \geq 1$ there is an $(n-1)$-connected n-dimensional simplicial complex X such that G acts on X with finite stabilizers and with an n-simplex as fundamental domain.*

The term "fundamental domain" here is to be understood in the strong sense: There is a closed n-simplex Δ which maps homeomorphically onto the quotient space X/G. Equivalently, every simplex of X is in the G-orbit of a unique face of Δ.

This theorem yields much more homological information about G than I was able to obtain in [4]. An immediate consequence, for instance, is that G is $\mathbf{Q}$-acyclic. The theorem also yields interesting combinatorial information about G: Take $n = 2$; then X is simply-connected, and it follows that G is the direct limit of the diagram

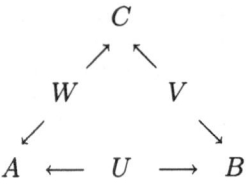

formed by the stabilizers of the vertices and edges of Δ, where all maps are inclusions. Equivalently, G is the free product of the vertex stabilizers A, B, C, amalgamated along their intersections.

This is interesting for two reasons. First, it settles an old question of Neumann and Neumann [8] about embeddings of finite amalgams. See §6 below for details. Secondly, it gives examples to illustrate some geometric notions recently introduced by Gersten and Stallings for "triangles of groups". In particular, the triangles that arise in the present paper are "positively curved", and they behave quite differently from the triangles of "non-positive curvature" studied by Gersten and Stallings.

For simplicity, I will prove the theorem and corollaries stated above only for Thompson's original group G rather than for an arbitrary $G \in \mathcal{G}$. The interested reader can easily generalize the proofs; this requires more complicated notation, but no new ideas.

The paper is organized as follows. §1 contains a brief treatment of triangles of groups, in order to provide the background for our later combinatorial study of Thompson's group G. The definition of G, as an algebraic automorphism group, is then reviewed in §2. This is followed by an optional §3, in which the same group is described as a group of homeomorphisms of the Cantor set. Readers are free to adopt, for the remainder of the paper, whichever definition they prefer. §4 is devoted to an unpublished result of Melanie Stein, which yields a contractible complex X on which G acts with finite stabilizers. It is a subcomplex of a contractible complex that was used in [4]. The main theorem is then proved in §5; the complexes called X in the statement of that theorem are obtained by suitably "truncating" the X of §4. Finally, §§6 and 7 contain corollaries: Combinatorial results are given in §6 and homological results in §7.

I am grateful to G. Higman for telling me about the problem of Neumann and Neumann and pointing out that my results solved it.

1. Triangles of Groups

Recall that the amalgamated free product $G = A *_U B$ is defined whenever one is given a diagram

$$A \leftarrow U \rightarrow B$$

of groups and monomorphisms. By definition, G is the direct limit of this diagram. It is well-known that the canonical maps $A \rightarrow G$ and $B \rightarrow G$ are injective. Moreover, if we identify A, B, and U with their images in G, then $U = A \cap B$.

In 1948 Hanna Neumann [9] considered a more general situation, where one is given an arbitrary family of groups $A, B, C, \ldots$ and, between any two of them, a common subgroup to amalgamate. We will be interested in the case of three free factors, in which case what we are given is a diagram

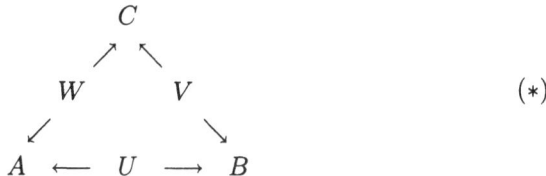

$$(*)$$

of groups and monomorphisms. We call such a diagram a *triangle of groups*, and we call A, B, C (resp. U, V, W) the *vertex groups* (resp. *edge groups*) of the triangle. Let G be the direct limit of the diagram $(*)$. Unlike the situation with two free factors, the canonical maps from the vertex groups to G need not be injective. But if they happen to be injective, and if in addition the image of each edge group in G is the intersection of the images of the corresponding vertex groups, then we call the triangle *realizable*. In this case G is said to be the *free product of A, B, and C, amalgamated along U, V, and W*, and we identify all the groups in $(*)$ with their images in G. Following Neumann [9], one sometimes uses the term "generalized amalgamated free product", to avoid confusion with the classical case where there are only two free factors.

REMARK. Readers familiar with Bass–Serre theory [11] will note that $(*)$ can be viewed as a graph of groups (where the underlying graph is the boundary of a 2-simplex). One is therefore tempted to form the fundamental group $\tilde{G}$ of this graph of groups. It is well-defined after a choice of basepoint or maximal tree. One such choice yields the following description of $\tilde{G}$: First construct $H = A *_U B *_V C$; then $\tilde{G}$ is the HNN extension $\langle H, t \; ; \; t^{-1}W't = W'' \rangle$, where W' (resp. W'') is the image of W in A (resp. C). Thus the direct limit G that we are studying is the quotient of $\tilde{G}$ obtained by introducing the relation $t = 1$.

Call the triangle (∗) *fillable* if it can be completed to a commutative diagram

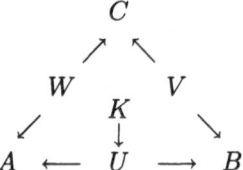

in which all maps are monomorphisms and each of the three squares has the following property: The image of K in the vertex group is the intersection of the images of the two edge groups. As Neumann [9] pointed out, fillability is a necessary condition for realizability; for if (∗) is realizable, then we can fill the triangle with $K = A \cap B \cap C$. But there are examples in [9] which show that fillability is not sufficient for realizability.

We turn now to connections with topology. Recall that Serre [11] has given a topological interpretation of ordinary amalgamated free products, in terms of group actions on trees with a 1-simplex as fundamental domain. There is an analogous result, due to Soulé [12] and Behr [1], for generalized amalgamated free products. In the case at hand, where there are three factors, the result is that amalgamated free product decompositions correspond to group actions on 1-connected 2-dimensional simplicial complexes with a 2-simplex as fundamental domain. More precisely, suppose G is the amalgamated free product associated to a realizable triangle (∗) as above, and let X be the (essentially unique) simplicial 2-complex such that G acts on X with a 2-simplex Δ as fundamental domain and with A, B, and C as the stabilizers of the vertices of Δ. Thus the vertex set of X is $G/A \amalg G/B \amalg G/C$, and a collection of vertices is a simplex if and only if the corresponding cosets have a non-empty intersection in G. Then X is 1-connected ([1], Satz 1.2). Conversely, if a group G acts on a 1-connected simplicial 2-complex with a 2-simplex Δ as fundamental domain, then G is the free product of the stabilizers of the vertices of Δ, amalgamated along their intersections (cf. [12] or [3]).

We conclude this section by stating some results of Gersten and Stallings [unpublished] concerning triangles of groups with "non-positive curvature". Gersten and Stallings begin by assigning an angle to each of the corners of the triangle (∗). To define the angle at A, for instance, let Y be the (essentially unique) 1-dimensional simplicial complex such that A acts on Y with an edge e as fundamental domain and with U and W as the stabilizers of the vertices of e. Let k be the smallest integer ≥ 3 such that Y contains a k-gon, if such an integer exists; otherwise, let $k = \infty$. Then the angle at A is defined to be $2\pi/k$. [The motivation for this comes from the fact that if the triangle is realizable and X is the associated 2-complex, then Y is the link in X of the vertex of Δ corresponding to A.]

The results of Gersten and Stallings concern triangles which are fillable and have angle sum $\leq \pi$. Their first theorem is that such a triangle is always realizable. They go on to prove a number of results about the generalized amalgamated free product G and the associated 2-complex X. For example, they show that X is contractible. In fact, they endow X with a geometric structure, making it, in some sense, a space of non-positive curvature in which any two points can be joined by a unique geodesic. They then use this geometric structure to prove that every bounded subgroup of G is conjugate to a subgroup of one of the vertex groups A, B, C. "Bounded" here means that there is an integer n such that every element of the given subgroup can be expressed as a product $x_1 \cdots x_l$ with $l \leq n$ and $x_i \in A \cup B \cup C$.

Suppose we add the hypothesis that the vertex groups are finite. Then the previous paragraph yields, among other things, the following two results: (a) G has only finitely many conjugacy classes of finite subgroups. (b) Any torsion-free subgroup of G has cohomological dimension at most 2.

We proceed now to Thompson's group. As we will see, it is the generalized amalgamated free product associated to a realizable triangle of finite groups. But Thompson's group has finite subgroups of arbitrarily large order, and it has torsion-free subgroups of infinite cohomological dimension. Our triangle, then, will necessarily have angle sum greater than π.

2. Thompson's Group

We begin with a very brief review of the definition of Thompson's group G as an algebraic automorphism group. See [6] and §4 of [4] for more details, further references, and historical remarks. See also §3 below for an alternative treatment, where G is viewed as a group of homeomorphisms of the Cantor set.

Consider the algebraic system consisting of a set S together with a bijection $\alpha : S \to S \times S$. For lack of a better name, we will simply call S an *algebra* in what follows. Let α_0 and α_1 be the components of α. They are unary operators $S \to S$, which we write on the right. Thus α is given by $x \mapsto (x\alpha_0, x\alpha_1)$. As an aid to the intuition, we will sometimes call $x\alpha_0$ and $x\alpha_1$ the two *halves* of x.

Let S be the free algebra on a single generator x. Then S has bases of arbitrary finite cardinality. For example, it has a 2-element basis $\{x\alpha_0, x\alpha_1\}$ and a 3-element basis $\{x\alpha_0\alpha_0, x\alpha_0\alpha_1, x\alpha_1\}$. These examples illustrate a general method for constructing bases: Given an r-element basis L and an element $y \in L$, we get an $(r+1)$-element basis M by replacing y by its two halves y_0, y_1. One says that M is a *simple expansion* of L. Conversely, if we start with a basis M and an ordered pair (y_0, y_1) of distinct elements of M, then we obtain a basis L by replacing y_0 and y_1 by $y = \alpha^{-1}(y_0, y_1)$. The basis L is called a *simple contraction* of M. Given two bases L, M, then, M is a simple expansion of L if and only if L is a simple contraction of M.

We now iterate these constructions: We say that M is an *expansion* of L and that L is a *contraction* of M if there is a sequence $L = L_0, \ldots, L_d = M$ $(d \geq 0)$ with L_i a simple expansion of L_{i-1} for $1 \leq i \leq d$. It is shown in [6] that any two bases have a common expansion. In other words, if L and M are arbitrary bases, then one can get from L to M by doing an expansion followed by a contraction. In particular, any basis can be constructed as a contraction of an expansion of the original basis $\{x\}$. Note, in this connection, that expansions of $\{x\}$ are very concrete objects; namely, they correspond to finite, rooted, binary trees. For example, the 3-element basis given in the previous paragraph is represented by the tree

Thompson's group G is now defined to be the group of automorphisms of S. This is the group called $G_{2,1}$ in [6] and [4]. To construct examples of elements of G, choose two expansions L, M of $\{x\}$ with the same cardinality, and choose an arbitrary bijection $L \to M$; then this bijection extends uniquely to an automorphism g of S. Moreover, every $g \in G$ can be described in this way.

By taking $L = M$ above, we see that G contains, for every positive integer n, the symmetric group on n letters. Consequently, every finite group can be embedded in G. In fact, every countable locally finite group can be embedded in G ([6], Theorem 6.6).

On the other hand, G also contains interesting torsion-free subgroups. For example, it contains the group F studied in [5] (and called $F_{2,1}$ in [4]). In particular, G contains torsion-free subgroups of infinite cohomological dimension.

3. Dyadic Homeomorphisms of the Cantor Set

This section, which may be omitted, presents an alternative view of Thompson's group.

Let C be the Cantor set of infinite sequences $a = (a_j)_{j \geq 1}$ with $a_j \in \{0, 1\}$. It is topologized as the product of infinitely many discrete two-point spaces. Note that there is a decomposition $C = C_0 \amalg C_1$, where C_i is the set of sequences a with $a_1 = i$. Each C_i is canonically homeomorphic to C, so we can iterate the process, thereby expressing C as the union of n copies of itself for any positive integer n. For example, we can partition C_0 into two Cantor sets C_{00} and C_{01} to get

$$C = C_{00} \amalg C_{01} \amalg C_1.$$

Finite partitions of C obtained in this way will be called *standard partitions*. There is one such partition for every finite, rooted, binary tree. Thus standard partitions correspond to expansions of $\{x\}$ in the notation of §2. The number of subspaces occurring in a partition will be called its *rank*. This is the same as the cardinality of the associated basis of S. It is also the same as the number of leaves of the associated tree, where a leaf is a node with no descendants.

Suppose now that we are given two standard partitions $L = (L_i)_{i \in I}$ and $M = (M_j)_{j \in J}$ with the same rank. For example, we could take $L = \{C_{00}, C_{01}, C_1\}$ and $M = \{C_0, C_{10}, C_{11}\}$. Choose a bijection between I and J. Then we can construct a homeomorphism $g : C \to C$ which maps each L_i to the corresponding M_j by the canonical homeomorphism. This makes sense because each L_i is canonically homeomorphic to C, as is each M_j. Any g constructed in this way will be called a *dyadic homeomorphism* of C.

The dyadic homeomorphisms form a group under composition, and this group G is Thompson's finitely presented infinite simple group, as described for instance in [7]. It is not hard to see that G is isomorphic to the group called G in §2, but we will not make any use of this isomorphism.

We close this section by discussing partitions of C more general than the standard ones. These play the role of bases of S more general than those which are expansions of $\{x\}$. We will need a precise definition of the word "partition", which we have been using informally up to now: A *partition* of C of rank n is a pair (L, H), where L is a collection of n pairwise disjoint subspaces of C whose union is C, and H is a collection of homeomorphisms $h : C \to D$, one for each $D \in L$. Roughly speaking, then, a partition exhibits C as the disjoint union of finitely many copies of itself. We will often suppress H from the notation and simply say that L is a partition.

Note that the standard partitions discussed earlier yield partitions in the present sense, since each subspace occurring in a standard partition is canonically homeomorphic to C. We wish to extend to arbitrary partitions the subdivision process used earlier to construct standard partitions.

Given a partition L and an element $D \in L$, we can write $D = hC_0 \amalg hC_1$, where $h : C \to D$ is the homeomorphism associated to D. Then each hC_i is canonically homeomorphic to C, so we obtain a new partition M by replacing D by its two "halves" hC_0 and hC_1. We say that M is a *simple expansion* of L. Conversely, given a partition M and an ordered pair (D_0, D_1) of distinct elements of M, we can construct a *simple contraction* L of M by replacing D_0 and D_1 by $D = D_0 \amalg D_1$, which is canonically homeomorphic to C. As in §2, we can iterate these constructions to obtain more general notions of expansion and contraction.

In contrast to the situation of §2, two partitions need not have a common expansion. We remedy this by restricting attention to a subset of the set of all partitions: A partition is called *admissible* if it is a contraction of some standard partition. It is easy to check that any two admissible partitions have a common expansion. The interested reader can now construct a bijection between the set of admissible partitions of C and the set of bases of S.

Finally, we wish to define an action of G on the set of admissible partitions. Note first that the full homeomorphism group of C acts on the set of all partitions: Given a partition (L, H) and a homeomorphism $g : C \to C$, there is a new partition (gL, gH), where $gL = \{gD : D \in L\}$ and

$gH = \{ gh : h \in H \}$. Here gh is to be interpreted as the composite

$$C \xrightarrow{h} D \xrightarrow{g|D} gD.$$

This action has the property that the stabilizer of a rank n partition is isomorphic to the symmetric group on n letters.

One can now verify:

PROPOSITION. *A partition is admissible if and only if it is in the G-orbit of a standard partition. In particular, the set of admissible partitions is invariant under the action of G.*   □

4. A contractible space for G

Our starting point is the set $\mathcal{B}$ of bases of S. [Readers of §3 may, if they wish, work instead with the set of admissible partitions of the Cantor set.] As in [4], we view $\mathcal{B}$ as a poset, where $L \leq M$ if M is an expansion of L. This poset structure is compatible with the obvious action of G on $\mathcal{B}$, i.e., G acts by poset automorphisms. Let $|\mathcal{B}|$ be the simplicial complex associated to $\mathcal{B}$; its simplices are the finite chains $L_0 < \cdots < L_n$ in $\mathcal{B}$. This is the G-complex with finite stabilizers which was used in [4] to study G. Since $\mathcal{B}$ is a directed set, $|\mathcal{B}|$ is contractible. The purpose of the present section is to present a result of Melanie Stein [unpublished], which provides a contractible G-invariant subcomplex of $|\mathcal{B}|$.

Given a basis L of S, an *elementary expansion* of L is a basis M obtained by choosing a subset $L' \subseteq L$ and replacing each $y \in L'$ by its two halves y_0, y_1. Thus we can get from L to M by a sequence of simple expansions, where at each stage we expand some $y \in L$ rather than one of the halves introduced by an earlier expansion. We write $L \preceq M$ if M is an elementary expansion of L, and we write $L \prec M$ if, in addition, $L \neq M$. [Warning: The relations "$\preceq$" and "$\prec$" are not transitive.]

Call a simplex $L_0 < \cdots < L_n$ *elementary* if $L_0 \preceq L_n$. This implies that $L_i \preceq L_j$ for $i \leq j$. Hence any face of an elementary simplex is elementary, and the elementary simplices form a (G-invariant) subcomplex $X \subset |\mathcal{B}|$.

THEOREM 1 (M. STEIN). *The complex X of elementary simplices is contractible.*

We will use, in the proof, the standard notation for intervals in a poset. For example, the open interval (L, M) is defined by

$$(L, M) = \{ N \in \mathcal{B} : L < N < M \}.$$

The closed interval $[L, M]$ and the half-open intervals $[L, M)$ and $(L, M]$ are defined similarly. The following lemma is the key step in the proof.

LEMMA. *If M is a non-elementary expansion of L, then $|(L, M)|$ is contractible.*

PROOF: For any expansion N of L, let N_0 be the largest element of $[L, N]$ such that $L \preceq N_0$; it is obtained by taking L', in the definition of "elementary expansion", to consist of all elements $y \in L$ which get expanded in the passage from L to N. Note that we have $N_0 \in (L, M)$ for any $N \in (L, M]$. The inequalities $N \geq N_0 \leq M_0$ now yield a "conical" contraction of (L, M), cf. [10], 1.5. □

Note that the complexes $|[L, M]|$, $|[L, M)|$, and $|(L, M]|$ are also contractible, for trivial reasons, since the intervals in question all have a largest or smallest element.

PROOF OF THEOREM 1: Since $|\mathcal{B}|$ is already known to be contractible, it suffices to show that we can obtain $|\mathcal{B}|$ from X by a sequence of adjunctions which do not change the homotopy type. To this end, we construct $|\mathcal{B}|$ by successively adjoining the subcomplexes $|[L, M]|$ with M a non-elementary expansion of L. We do these adjunctions by induction on $r(M) - r(L)$, where $r(\ )$ denotes the cardinality of a basis. Thus the part of $|[L, M]|$ already present at the time of the adjunction is $|[L, M) \cup (L, M]|$, which is the suspension of $|(L, M)|$ and hence is contractible by the lemma. Since $|[L, M]|$ is also contractible, we conclude that the adjunction has no effect on the homotopy type. □

REMARK. Everything in this section goes through without change if we fix an integer r and replace $\mathcal{B}$ by the set of bases of cardinality $\geq r$.

5. Proof of the Main Theorem

Recall from the introduction that our goal is to construct for any $n \geq 1$ an $(n - 1)$-connected n-dimensional simplicial complex on which G acts with finite stabilizers and with an n-simplex as fundamental domain. We will do this by "truncating" the complex X of elementary simplices. Given integers p, q with $1 \leq p \leq q$, let $X_{p,q}$ be the full subcomplex of X generated by the vertices L with $p \leq r(L) \leq q$. Here, as above, $r(\ )$ denotes the cardinality of a basis. Note that $X_{p,q}$ is G-invariant. Note also that the dimension of $X_{p,q}$ satisfies $\dim X_{p,q} \leq q - p$, with equality if $q \leq 2p$.

Now fix an integer $n \geq 1$, and consider the G-complex $X_{p,p+n}$ for $p \geq n$. It is n-dimensional. Associated to any simplex $L_0 < \cdots < L_m$ of this complex is a sequence of integers $r(L_0) < \cdots < r(L_m)$, called the *type* of the simplex. It is easy to check that two simplices are G-equivalent if and only if they have the same type. Since every n-simplex has exactly one face of each possible type, we conclude that the action of G on $X_{p,p+n}$ admits an n-simplex as fundamental domain. The main theorem now follows from:

THEOREM 2. *There is an integer p_0 (depending on n) such that $X_{p,p+n}$ is $(n-1)$-connected for $p \geq p_0$.*

The proof will make use of a family of simplicial complexes K_r ($r \geq 2$), defined as follows: The vertices of K_r are the ordered pairs (a,b) with $a, b \in \{1, \ldots, r\}$ and $a \neq b$; a collection $\{(a_0, b_0), \ldots, (a_m, b_m)\}$ of such vertices is a simplex if $\{a_i, b_i\} \cap \{a_j, b_j\} = \emptyset$ for $i \neq j$. The complexes K_r appeared in [4], where it was shown that the connectivity of K_r tends to ∞ with r (cf. [4], Lemma 4.20). In other words:

LEMMA. *There is an integer r_0 (depending on n) such that K_r is $(n-1)$-connected for $r \geq r_0$.* $\qquad\square$

PROOF OF THEOREM 2: Fix p and consider the sequence of inclusions

$$X_{p,p+n} \subset X_{p,p+n+1} \subset X_{p,p+n+2} \subset \cdots.$$

The union is the full subcomplex of X generated by the vertices L with $r(L) \geq p$. This union is contractible by the remark at the end of §4. So $X_{p,p+n}$ will be $(n-1)$-connected if the inclusion $X_{p,q} \hookrightarrow X_{p,q+1}$ induces an isomorphism $\pi_i(X_{p,q}) \xrightarrow{\approx} \pi(X_{p,q+1})$ for all $i < n$ and all $q \geq p+n$. Now this inclusion is obtained by adjoining, for each basis M with $r(M) = q+1$, a cone over the link of M in $X_{p,q}$. Fix M and let Y be this link. Its vertices are the bases L with $L \prec M$ and $r(L) \geq p$, and its simplices are the chains $L_0 < \cdots < L_m$ of such bases.

We can describe a basis $L \prec M$ by specifying the pairs of elements of M which are contracted to get L. Thus Y has one vertex for every non-empty set $P \subset M \times M$ satisfying:

(1) If $(a,b) \in P$, then $a \neq b$.
(2) $\{a,b\} \cap \{c,d\} = \emptyset$ for any two distinct pairs $(a,b), (c,d) \in P$.
(3) P has cardinality $\leq q + 1 - p$.

And the simplices of Y correspond to chains $P_0 \subset \cdots \subset P_m$ of such sets P. Now sets P satisfying (1) and (2) are the same as simplices of the complex K_{q+1} defined above. And condition (3) says that the simplex has dimension at most $q - p$. So our complex Y, which consists of chains of such simplices, is the barycentric subdivision of the $(q - p)$-skeleton of K_{q+1}. The lemma now implies that Y is $(n - 1)$-connected if $q + 1 \geq r_0$ and $q - p \geq n$.

This last inequality is vacuous, since we are only considering integers $q \geq p + n$. And the first will be satisfied for all $q \geq p + n$ provided $p + n + 1 \geq r_0$. So if we take $p_0 = r_0 - n - 1$ and $p \geq p_0$, then the complexes Y that arise above will all be $(n - 1)$-connected. Attaching a cone over such a Y does not affect π_i for $i < n$; hence $X_{p,p+n}$ is $(n - 1)$-connected, as required. $\square$

REMARK. K. Vogtmann [private communication] has shown that one can take $r_0 = 3n + 2$, which gives $p_0 = 2n + 1$. Thus $X_{3,4}$ is connected, $X_{5,7}$ is 1-connected, $X_{7,10}$ is 2-connected, etc. Here is a proof of Vogtmann's result for the case $n = 2$:

We must show that K_r is 1-connected for $r \geq 8$. It is trivial to verify that K_r is connected, so the content of the assertion is that any closed edge path is null-homotopic. To prove this, it suffices to show that any edge path of length 3 is homotopic, relative to its endpoints, to a shorter path. Let v_0, v_1, v_2, v_3 be the vertices of an edge path of length 3. If v_0 and v_2 are joined by an edge in K_r (i.e., if the corresponding 2-element subsets of $\{1, \ldots, r\}$ are disjoint), then the first three vertices of our path are the vertices of a 2-simplex, and the path can be shortened to v_0, v_2, v_3. A similar remark applies to v_1 and v_3. So we may assume that v_0 and v_2 have an element of $\{1, \ldots, r\}$ in common, and similarly for v_1 and v_3. But then the entire edge path involves at most 6 elements of $\{1, \ldots, r\}$. Since $r \geq 8$, we can now find a vertex v such that the given path is in the star of v, hence it is homotopic to v_0, v, v_3.

6. A Combinatorial Description of G

We specialize now to the case $n = 2$. As we have just proved, the 2-complex $X_{p,p+2}$ is 1-connected for any $p \geq 5$, and G acts on it with a 2-simplex Δ as fundamental domain. Take $p = 5$, for instance. Then the vertices of Δ can be written in the form

$$L_5 = \{a, b, c, d, e\}$$
$$L_6 = \{a, b, c, d, e_0, e_1\}$$
$$L_7 = \{a, b, c, d_0, d_1, e_0, e_1\},$$

where $d_i = d\alpha_i$ and $e_i = e\alpha_i$. The stabilizers of these vertices are iso-morphic to Σ_5, Σ_6, and Σ_7, respectively, where Σ_r is the symmetric group on r letters. It is a simple matter to work out the intersections of these stabilizers. The results stated in §1 now yield:

THEOREM 3. *There is a realizable triangle of groups*

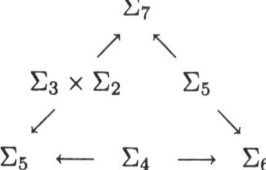

whose amalgamated free product is Thompson's infinite simple group. □

The maps in this triangle can be described as follows. Regard Σ_r as the group of permutations of $\{1, \ldots, r\}$. The maps $\Sigma_4 \hookrightarrow \Sigma_r$ $(r = 5, 6)$ are then the standard inclusions, obtained by letting Σ_4 permute $\{1, 2, 3, 4\} \subset \{1, \ldots, r\}$. Similarly, the maps $\Sigma_5 \hookrightarrow \Sigma_r$ $(r = 6, 7)$ are obtained by letting Σ_5 permute $\{1, 2, 3, r-1, r\}$. Finally, $\Sigma_3 \times \Sigma_2$ is embedded in Σ_5 in the standard way, with the first factor acting on $\{1, 2, 3\}$ and the second factor acting on $\{4, 5\}$; but the embedding $\Sigma_3 \times \Sigma_2 \hookrightarrow \Sigma_7$ is not of the standard type. The first factor permutes $\{1, 2, 3\}$, but the non-trivial element of the second factor, instead of mapping to a transposition, maps to the product $(46)(57)$ of two transpositions.

REMARK. It is easy to describe links of vertices in $X_{5,7}$, from which one can compute the Gersten–Stallings angles of the triangle above (cf. §1). These angles turn out to be $\pi/3$, $\pi/2$, and $\pi/3$ at the vertices stabilized by Σ_5, Σ_6, and Σ_7, respectively. Note that the sum of these angles is greater than π.

Finally, we explain how Theorem 3 settles a question of Neumann and Neumann [8]. Let A, B, C denote the vertex groups in the triangle above, and let H be their union $A \cup B \cup C$ in G. We view H as a set with a partially defined multiplication, where the product of two elements is defined if at least one of the groups A, B, C contains both of them. In the language of [8], H is an *amalgam* of the groups A, B, C. In general, an amalgam of groups need not be embeddable in a group; but our amalgam is embedded in Thompson's group, by construction.

The question posed in [8], p. 255, is the following: If an amalgam of groups is finite and is embeddable in a group, is it necessarily embeddable in a finite group? The following corollary of Theorem 3 shows that the answer is "no":

COROLLARY. *The finite amalgam H is not embeddable in a finite group.*

PROOF: Since G is the amalgamated free product of A, B, C, it follows that the inclusion $H \hookrightarrow G$ is the universal homomorphism from H to a group. So if H could be embedded in a finite group, then it would embed in a finite quotient of G. But G is infinite and simple, so its only finite quotient is the trivial group.   □

Note that this is not just an isolated example. For we could replace 5 by any $p \geq 5$ above, thereby getting infinitely many triangles with similar properties. Moreover, as stated in the introduction, we could replace G by any member of Higman's family of finitely presented infinite simple groups.

7. Homology Calculations

THEOREM 4. *Thompson's group G is $\mathbf{Q}$-acyclic, i.e., $H_i(G, \mathbf{Q}) = 0$ for all $i > 0$.*

PROOF: Fix n, and let $Y = X_{p,p+n}$ for large p. Consider the equivariant homology $H_*^G(Y, \mathbf{Q})$, as defined for instance in [2], §VII.7. Since the stabilizer of every simplex of Y is finite, this homology is isomorphic to $H_*(Y/G)$ (cf. [2], §VII.7, Exercise 2). But Y/G is a simplex, so we have $H_i^G(Y, \mathbf{Q}) = 0$ for $i > 0$. On the other hand, Y is $(n-1)$-connected, so the canonical map $H_i^G(Y) \to H_i(G)$ is an isomorphism for $i < n$ and an epimorphism for $i = n$ (cf. [2], VII.7.2). Hence $H_i(G, \mathbf{Q}) = 0$ for $i \leq n$. Since n is arbitrary, this completes the proof.   □

REMARK. There is some evidence to suggest that G might be $\mathbf{Z}$-acyclic. For example, I have proven the following:

(1) $H_i(G) = 0$ for $i = 1, 2, 3$. In particular, the Schur multiplier of G is trivial.

(2) Let Σ_r be embedded in G as the stabilizer of an r-element basis, and let k be any field. Then the induced map $H_i(\Sigma_r, k) \to H_i(G, k)$ is the zero map for all $i > 0$.

(3) Let F be the torsion-free subgroup of G mentioned at the end of §2. Then the inclusion $F \hookrightarrow G$ induces the zero map $H_i(F) \to H_i(G)$ for all $i > 0$.

Assertion (1) is proved by spectral sequence computations. The proofs of (2) and (3), however, are more interesting; they are based on the fact that there is an embedding $G \times G \hookrightarrow G$, which induces a ring structure on $H_*(G)$. Details will be given elsewhere.

REFERENCES

1. H. Behr, *Explizite Präsentationen von Chevalleygruppen über Z*, Math. Z. **141** (1975), 235–241.
2. K. S. Brown, "Cohomology of Groups," Graduate Texts in Mathematics 87, Springer-Verlag, New York, 1982.
3. K. S. Brown, *Presentations for groups acting on simply-connected complexes*, J. Pure Appl. Algebra **32** (1984), 1–10.
4. K. S. Brown, *Finiteness properties of groups*, J. Pure Appl. Algebra **44** (1987), 45–75.
5. K. S. Brown and R. Geoghegan, *An infinite-dimensional torsion-free FP_∞ group*, Invent. Math. **77** (1984), 367–381.
6. G. Higman, "Finitely Presented Infinite Simple Groups," Notes on Pure Mathematics 8, Australian National University, Canberra, 1974.
7. R. McKenzie and R. J. Thompson, *An elementary construction of unsolvable word problems in group theory*, in "Word Problems," Proc. Conf. Irvine 1969 (edited by W. W. Boone, F. B. Cannonito, and R. C. Lyndon), Studies in Logic and the Foundations of Mathematics, vol. 71, North-Holland, Amsterdam, 1973, pp. 457-478.
8. B. H. Neumann and H. Neumann, *A contribution to the embedding theory of group amalgams*, Proc. London Math. Soc. (3) **3** (1953), 243–256. ("Selected Works of B. H. Neumann and Hanna Neumann," vol. III, Charles Babbage Research Centre, Winnipeg, 1988, pp. 439–452.)
9. H. Neumann, *Generalized free products with amalgamated subgroups*, Amer. J. Math. **70** (1948), 590–625.
10. D. Quillen, *Homotopy properties of the poset of non-trivial p-subgroups of a group*, Advances in Math. **28** (1978), 101–128.
11. J-P. Serre, "Trees," Springer-Verlag, Berlin, 1980. Translation of "Arbres, Amalgames, SL_2", Astérisque **46**, 1977.
12. C. Soulé, *Groupes opérant sur un complexe simplicial avec domaine fondamental*, C. R. Acad. Sci. Paris Sér. A **276** (1973), 607–609.

Mathematics Department, White Hall, Cornell University, Ithaca, NY 14853

Research at MSRI partially supported by a grant from the National Science Foundation.

The Geometry of Rewriting Systems: A Proof of the Anick–Groves–Squier Theorem

KENNETH S. BROWN

Abstract. Let G be a group or monoid which is presented by means of a complete rewriting system. Then one can use the resulting normal forms to collapse the classifying space of G down to a quotient complex (typically "small") of the same homotopy type. If the rewriting system is finite, then the quotient complex has only finitely many cells in each dimension. The proof yields an explicit free resolution of $\mathbf{Z}$ over $\mathbf{Z}G$, similar to resolutions obtained by Anick, Groves, and Squier.

Introduction

Several years ago Ross Geoghegan and I [6] were interested in the homological finiteness properties of a certain group G. We succeeded in constructing a small $K(G,1)$-complex, with only two n-cells for each $n \geq 1$, by an indirect method: We first built a big $K(G,1)$, with infinitely many n-cells for each n, and we then "collapsed away" all but two cells in each dimension. This notion of "collapse" will be explained in §1; for now, one can think of it as analogous to collapsing a maximal tree in a connected complex in order to get rid of all the vertices but one.

The method seemed ad hoc at the time, but it turns out to have much wider applicability than I would have guessed. In particular, it applies to groups and other algebraic objects which come equipped with a complete rewriting system [definition in §2 below]. One simply uses the classical bar construction as the starting point, i.e., as the "big" complex on which to perform the collapse, and one uses the normal forms that come from the rewriting system in order to figure out which cells to collapse.

The result of this process, in the case of a group G, is an explicit $K(G,1)$-complex, typically much smaller than the classical one. One also obtains an explicit free resolution of $\mathbf{Z}$ over $\mathbf{Z}G$, similar to resolutions obtained by Anick [1], Groves [11], and (through dimension 3) Squier [19]. In particular, we recover the Anick–Groves–Squier result that G is of type FP_∞ if the rewriting system is finite.

We begin by explaining in §1 the collapsing method of [6] that we will be using. In §2 we review rewriting systems. The application of the collapsing method to monoids with a rewriting system is then given in §3. We translate the result into algebraic language in §4 and show how it leads to a free resolution. It is easy to describe the basis elements of this resolution, but it is not so clear how to compute the boundary operator. This question is treated in §§5 and 6. Finally, §7 contains some examples.

I am grateful to D. E. Cohen and the referee for helpful suggestions that improved my exposition.

1. Collapsing BM: An Overview

Recall that a monoid M gives rise to a semi-simplicial complex BM, whose n-simplices are n-tuples $\sigma = (m_1, \ldots, m_n)$ of elements of M. The face operators are given by

$$d_i\sigma = \begin{cases} (m_2, \ldots, m_n) & i = 0 \\ (m_1, \ldots, m_{i-1}, m_i m_{i+1}, m_{i+2}, \ldots, m_n) & 0 < i < n \\ (m_1, \ldots, m_{n-1}) & i = n, \end{cases}$$

and the degeneracy operators are given by

$$s_i\sigma = (m_1, \ldots, m_i, 1, m_{i+1}, \ldots, m_n) \qquad (0 \le i \le n).$$

The geometric realization $X = |BM|$ is called the *classifying space* of M. It is a CW-complex with one n-cell for every non-degenerate n-simplex of BM, i.e., for every n-tuple $(m_1, \ldots, m_n)$ with $m_i \ne 1$ for all i. We will use the same symbol $(m_1, \ldots, m_n)$ for both a simplex of BM and the corresponding cell of X.

A glance at the 2-skeleton of X shows that $\pi_1(X)$ is the group completion of M, i.e., it is the target of a monoid homomorphism $M \to \pi_1(X)$ which is universal for homomorphisms from M to a group. In particular, $\pi_1(X) = M$ if M is a group. Moreover, X is an Eilenberg–MacLane complex of type $K(M, 1)$ in this case; in fact, it is the "original" $K(M, 1)$ constructed by Eilenberg and MacLane [9] and, independently, by Eckmann [8]. If M is not a group, then X may or may not be an Eilenberg–MacLane complex (cf. [16], [10]).

Our method for analyzing the homotopy type of X is based on the following trivial observation. Suppose we are building X by attaching cells, one at a time, and we are ready to attach a 1-cell $\tau = (m)$. Suppose m admits a non-trivial factorization $m = m'm''$ such that the 1-cells (m') and (m'')

have already been adjoined. Then we can adjoin τ without changing the homotopy type, provided we simultaneously adjoin the 2-cell $\sigma = (m', m'')$. For we have $\tau = d_1\sigma$, and the other two faces of σ are already present; so the adjunction is an elementary expansion, and the resulting inclusion is a homotopy equivalence. It has a canonical homotopy inverse, which "collapses" σ onto the union of the two faces other than the "free face" τ.

In this situation we will say that τ is *redundant*, since we were able to adjoin it without changing the homotopy type. The chosen σ will be called the *collapsible* cell associated to τ and will be denoted $c(\tau)$. The construction has to start somewhere, of course, so we cannot expect all 1-cells to be redundant. Those that we start with will be called *essential*. In practice, they will be the cells (s), where s ranges over some set of generators of M. [Note: We use monoid generators, even if M happens to be a group.]

Turning now to the 2-skeleton, we have already adjoined some of the 2-cells, namely, the collapsible ones. We can expect some of the non-collapsible cells to be "essential", and we must simply adjoin them. In practice, these will correspond to some set of defining relations for M. The remaining 2-cells τ will be declared "redundant", and we will try to find for each such τ a "collapsible" 3-cell $\sigma = c(\tau)$, so that the adjunction of σ and τ can be done as an elementary expansion. Thus τ should be a face of σ, and all other faces should be present already when σ and τ are adjoined.

Continuing this process, we hope to classify the cells of X in all dimensions as "essential", "collapsible", or "redundant". We want to then build X in such a way that the redundant and collapsible cells can be adjoined without changing the homotopy type, so that X will be homotopy equivalent to a complex Y with one cell for each essential cell of X. Let's spell out explicitly what we need in order for this program to work. We will do this in the context of an arbitrary semi-simplicial complex.

Let K be a semi-simplicial complex and let X be its geometric realization. As above, we will identify the set of cells of X with the set of non-degenerate simplices of K. Assume that the cells have been partitioned into three classes, whose elements are called *essential, collapsible,* and *redundant*, respectively. The collapsible cells are required to have dimension ≥ 1. Assume further that to each redundant n-cell τ we have associated a collapsible $(n+1)$-cell $\sigma = c(\tau)$ and an integer $i = i(\tau)$ such that $\tau = d_i\sigma$. We will often refer to τ as the *free face* of σ. If τ' is a redundant n-cell such that $\tau' = d_j\sigma$ for some $j \neq i$, then we call τ' an *immediate predecessor* of τ and write $\tau' \prec \tau$. The point of this is that when we adjoin τ and σ, we want any immediate predecessor of τ to be present already.

The given cell partition and functions c and i will be said to constitute a *collapsing scheme* for K if the following two conditions are satisfied:

(C1) *The function c defines, for each $n \geq 0$, a bijection from the set of redundant n-cells to the set of collapsible $(n+1)$-cells.*

(C2) *There is no infinite descending chain $\tau \succ \tau' \succ \tau'' \succ \cdots$ of redundant n-cells.*

Note, as a special case of **(C2)**, that one cannot have $\tau \succ \tau$. In other words, there is a *unique* integer i such that $\tau = d_i c(\tau)$. Another consequence of **(C2)** is that, for any redundant cell τ, there cannot exist arbitrarily long descending chains $\tau = \tau_0 \succ \tau_1 \cdots \succ \tau_k$. This follows from König's lemma (cf. [13], §2.3.4.3), which is applicable since every redundant cell has only finitely many immediate predecessors. The maximal length k of a descending chain as above will be called the *height* of τ.

It is now a simple matter to achieve the goal stated above:

PROPOSITION 1. *Let K be a semi-simplicial complex with a collapsing scheme. Then its geometric realization $X = |K|$ admits a canonical quotient CW-complex Y, whose cells are in 1–1 correspondence with the essential cells of X. The quotient map $q : X \to Y$ is a homotopy equivalence. It maps each open essential cell of X homeomorphically onto the corresponding open cell of Y, and it maps each collapsible $(n+1)$-cell into the n-skeleton of Y.*

PROOF: Write X as the union of an increasing sequence of subcomplexes

$$X_0 \subseteq X_0^+ \subseteq X_1 \subseteq X_1^+ \subseteq \cdots,$$

where X_0 consists of the essential vertices, X_n^+ is obtained from X_n by adjoining the redundant n-cells and the collapsible $(n+1)$-cells, and X_{n+1} is obtained from X_n^+ by adjoining the essential $(n+1)$-cells. We can factor the inclusion $X_n \hookrightarrow X_n^+$ as a sequence of adjunctions

$$X_n = X_n^0 \subseteq X_n^1 \subseteq X_n^2 \subseteq \cdots,$$

where we construct X_n^{j+1} from X_n^j by adjoining τ and $c(\tau)$ for every redundant n-cell τ of height j. Note that every face of $c(\tau)$ other than τ is either an immediate predecessor of τ (and hence has height $< j$) or else is essential, collapsible, or degenerate. These faces are therefore already present, and the adjunction of τ and $c(\tau)$ is an elementary expansion.

[Note: The degenerate faces present no problem here since they are identified, when the geometric realization X is constructed, with cells of lower dimension. More precisely, let $\chi : \Delta^{n+1} \to X$ be the characteristic map for $\sigma = c(\tau)$, where Δ^{n+1} is the standard $(n+1)$-simplex. If some face of σ is degenerate, then χ maps the corresponding face of Δ^{n+1} into the $(n-1)$-skeleton of X. So χ maps all but the ith face of Δ^{n+1} into X_n^j, where $i = i(\tau)$, whence our assertion that the adjunction is an elementary expansion.]

The passage from X_n^j to X_n^{j+1} therefore consists of a possibly infinite number of simultaneous elementary expansions. In particular, we have a homotopy equivalence $X_n^j \hookrightarrow X_n^{j+1}$ for each j, whence a homotopy equivalence $X_n \hookrightarrow X_n^+$. Moreover, the collapsing maps associated to the elementary expansions above yield a canonical homotopy inverse $X_n^+ \to X_n$. The rest of the proof is an exercise in elementary homotopy theory. See [6], proof of Theorem 5.3, for more details. $\square$

When we apply these ideas to $K = BM$, we will construct $c(\tau)$ for a redundant cell $\tau = (m_1, \dots, m_n)$ by factoring one of the m_i, as we indicated above in the case $n = 1$. So we will need some reasonable way of factoring elements of M. This suggests that we need normal forms. It turns out that the normal forms coming from a complete rewriting system do the trick. The next section will be devoted to a review of rewriting systems, and the collapsing scheme will then be constructed in §3.

2. Rewriting Systems and Normal Forms

Let M be a monoid with a fixed set S of generators. Thus M is a quotient of the free monoid F on S. We will call the elements of S "letters" when we are thinking of S as a subset of F, and we will call the elements of F "words"; a word, then, can be viewed as a finite (possibly empty) string of letters. A subset $\mathcal{I} \subseteq F$ will be called a set of *normal forms* for M if $\mathcal{I}$ maps bijectively onto M under the quotient map $\pi : F \to M$. Such a set determines a function $r : F \to \mathcal{I}$ such that $r(w)$ is the unique element of $\mathcal{I}$ with $\pi(r(w)) = \pi(w)$. We are interested in normal forms with the property that $r(w)$ is computable from w by "rewriting", in a sense which we now explain.

Let R be a subset of $F \times F$ such that M admits the presentation $\langle\, S\,;\,R\,\rangle$. In other words, M is obtained from F by introducing one relation $w_1 = w_2$ for each $(w_1, w_2) \in R$. The elements of R will be called *rewriting rules*. It is important that they are *ordered* pairs of words. We will often emphasize this by writing $w_1 \to w_2$ if $(w_1, w_2) \in R$. More generally, we write $w \to w'$ whenever $w = u w_1 v$ and $w' = u w_2 v$ for some $(w_1, w_2) \in R$ and some $u, v \in F$. We then say that w' is obtained from w by *rewriting*, or *reduction*. A word w is called *reducible* if such a reduction is possible, and it is called *irreducible* otherwise.

We say that R is a *complete rewriting system* for M if it satisfies the following two conditions:

(R1) *The set $\mathcal{I}$ of irreducible words is a set of normal forms for M.*

(R2) *There is no infinite chain $w \to w' \to w'' \to \cdots$ of reductions.*

It follows from these axioms that we can compute $r(w)$ for $w \in F$ by starting with w and applying an arbitrary sequence of reductions, until we arrive at an irreducible word. Condition **(R2)** guarantees that this will eventually happen, and clearly the irreducible word we have reached is $r(w)$.

There are various ways of reformulating the axioms; see, for instance, [2], [14], or [19]. See also [12] for further information about rewriting systems.

Here are two simple examples. Several additional examples will be given in §7, and many further examples can be found in the references cited above, especially [14].

EXAMPLES. 1. Let M be the free commutative monoid generated by a set S. If we totally order S, then M admits as normal forms the set of words $s_1 \cdots s_n$ with $s_1 \le \cdots \le s_n$ in the chosen ordering. This is the set of irreducible words associated to a complete rewriting system with one rewriting rule $ts \to st$ for every pair $s, t \in S$ with $t > s$.

2. Let G be the one-relator group with two generators a, t and the defining relation $t^{-1} a t = a^2$. Let M be the submonoid of G generated by a and t. It is not hard to show that M admits the words $t^i a^j$ $(i, j \ge 0)$ as normal forms. (One can use, for instance, the fact that G is a semi-direct product $\mathbf{Z}[1/2] \rtimes \mathbf{Z}$.) This set of normal forms is the set of irreducible words associated to a complete rewriting system with one rewriting rule $at \to ta^2$.

(Exercise: Verify (**R2**). Some thought is required here, since application of the rewriting rule increases the length of a word.)

Assume now that we have a complete rewriting system. It will be convenient to introduce an ordering on words which reflects both the rewriting process and the subword relation. Recall first that a *subword* of a word w is any word w' such that $w = uw'v$ for some $u, v \in F$. If u and v are not both empty, then w' is a *proper* subword of w. We now denote by "$\prec$" the smallest transitive relation on F such that $w' \prec w$ if w' is a proper subword of w or if there is a reduction $w \to w'$. Explicitly, then, we have $w' \prec w$ if and only if either (i) w' is a proper subword of w or (ii) w' is a subword (not necessarily proper) of a word w'' such that there is a chain of reductions $w = w_0 \to \cdots \to w_k = w''$ for some $k \geq 1$.

The following assertion is an easy consequence of (**R2**):

(**R3**) *There is no infinite chain $w \succ w' \succ w'' \succ \cdots$ of words.*

For the purposes of this paper, it is not important that we have a particular rewriting system; the important thing for us, rather, is the set $\mathcal{I}$ of irreducible words. To emphasize this point of view, we will call a set of normal forms *good* if it is the set $\mathcal{I}$ of irreducible words associated to some complete rewriting system R for M.

REMARK. Squier [19] has noted that there is a canonical choice of complete rewriting system R for a given good set $\mathcal{I}$ of normal forms. Namely, let $\mathcal{L}$ be the set of words $w \notin \mathcal{I}$ such that every proper subword of w is in $\mathcal{I}$; then the canonical R consists of the rules $w \to r(w)$ for $w \in \mathcal{L}$.

Finally, we mention one further condition that we will sometimes assume is satisfied by R (or $\mathcal{I}$).

(**R4**) *Every $s \in S$, viewed as a word of length 1, is irreducible.*

This condition is harmless. For if it fails, then we can replace S by the subset $S' = S \cap \mathcal{I}$. This is still a set of generators of M, since $\mathcal{I}$ is contained in the submonoid $F' \subset F$ generated by S'. (To see this, note that $\mathcal{I}$ is closed under passage to subwords.) Moreover, $\mathcal{I}$ is a good set of normal forms satisfying (**R4**) with respect to the generating set S'.

3. Good Normal Forms Yield a Collapsing Scheme

Assume throughout this section that M is a monoid with a generating set S and a good set $\mathcal{I}$ of normal forms. As a reminder of what "good" means, we call the elements of $\mathcal{I}$ *irreducible* and the elements of $F - \mathcal{I}$ *reducible*. We are going to use $\mathcal{I}$ to construct a collapsing scheme for BM. The construction depends only on $\mathcal{I}$, and not on any particular choice of rewriting system R. But the verification of the crucial axiom **(C2)** will require a choice of R. The reader may prefer to simply assume from the outset that we are working with a specific R, e.g., the canonical one.

We will identify M, as a set, with $\mathcal{I}$. We must then be careful to distinguish between multiplication in the free monoid F, denoted by $(w_1, w_2) \mapsto w_1 w_2$, and multiplication in M, denoted by $(w_1, w_2) \mapsto w_1 * w_2$. The latter is defined for $w_1, w_2 \in \mathcal{I}$, and it is given by

$$w_1 * w_2 = r(w_1 w_2).$$

The simplices of our semi-simplicial complex BM are now viewed as n-tuples $(w_1, \ldots, w_n)$ of irreducible words, and the definition of the face operator d_i for $0 < i < n$ becomes

$$d_i(w_1, \ldots, w_n) = (w_1, \ldots, w_i * w_{i+1}, \ldots, w_n).$$

Finally, the cells of $X = |BM|$ are the n-tuples of *non-empty* irreducible words.

We now wish to carry out the procedure described in §1. Let's look first at low dimensions, for motivation. The unique vertex of X will be essential. A 1-cell $\tau = (w)$ is going to be called essential if $w \in S$ and redundant otherwise. If it is redundant, then the associated collapsible 2-cell will be $c(\tau) = (s, w')$, where $w = sw'$ with $s \in S$. Note that (s, w') is in fact a cell, because s and w' are non-empty subwords of w.

At this point we know that the collapsible 2-cells should be those of the form (s, w) with $s \in S$ and sw irreducible. The other 2-cells τ have the form (i) (w_1, w_2) with $w_1 \notin S$, or (ii) (s, w) with sw reducible. In case (i), τ will be called redundant, and we will set $c(\tau) = (s, w', w_2)$ $(s \in S, sw' = w_1)$. In case (ii), we would like to take $c(\tau) = (s, w', w'')$ for some factorization $w = w'w''$. Note, however, that we must choose this factorization so that sw' is reducible; for otherwise (s, w', w'') has already been used as $c(\tau')$, where τ' is the redundant cell (sw', w'') of type (i). Thus we are only going to be able to make (s, w) redundant if sw has a proper initial subword which is reducible. Those (s, w) for which this fails will be the essential 2-cells. Equivalently, (s, w) will be essential if and only if $sw \in \mathcal{L}$, where $\mathcal{L}$ is the set defined in the remark in §2 above.

(This is quite reasonable from the point of view of generators and relations. If (**R4**) holds, for instance, then there will be one essential 1-cell for each element of S and one essential 2-cell for each element of R, where R is the canonical set of rewriting rules for $\mathcal{I}$.)

These considerations lead to the following definitions. An n-cell $\tau = (w_1, \ldots, w_n)$ of X is called *essential* if it satisfies the following three conditions:

(1) $w_1 \in S$.

(2) $w_i w_{i+1}$ is reducible for $1 \leq i < n$.

(3) Every proper initial subword of $w_i w_{i+1}$ $(1 \leq i < n)$ is irreducible.

Note that it suffices, in verifying (3), to check that the subword of $w_i w_{i+1}$ obtained by deleting the last letter is irreducible.

If τ is not essential, let i be the largest integer such that the "$(i - 1)$-dimensional front face" $(w_1, \ldots, w_{i-1})$ of τ is an essential $(i - 1)$-cell. We have $1 \leq i \leq n$, where the case $i = 1$ occurs if the essential front face is empty, i.e., if w_1 has length $l(w_1) > 1$. If $i = 1$, then we call τ *redundant*, and we set $c(\tau) = (s, w', w_2, \ldots, w_n)$, where $w_1 = sw'$ with $s \in S$. Then $\tau = d_1 c(\tau)$, and we set $i(\tau) = 1$.

Suppose now that $i > 1$. Then either $w_{i-1} w_i$ is irreducible, in which case we say that τ is *collapsible*, or else some proper initial subword of $w_{i-1} w_i$ is reducible, in which case we call τ *redundant*. In the second case, write $w_i = w'w''$, where w' is the smallest initial subword of w_i such that $w_{i-1} w'$ is reducible. The words w' and w'' are necessarily non-empty and irreducible, and we set

$$c(\tau) = (w_1, \ldots, w_{i-1}, w', w'', w_{i+1}, \ldots, w_n).$$

We have $\tau = d_i c(\tau)$, and we set $i(\tau) = i$.

REMARK. To understand what an essential cell $\tau = (w_1, \ldots, w_n)$ looks like, consider the word $w = w_1 w_2 \cdots w_n$ obtained by ignoring the commas. It is not hard to show that τ is determined by w; in other words, there is only one way to re-insert commas into w so as to get an essential cell (cf. [1], Lemma 1.3). Note also that, by (2) and (3), some final subword of $w_i w_{i+1}$ is in $\mathcal{L}$ for $1 \leq i < n$. Consequently, the essential n-cells for $n \geq 3$ correspond to certain words w, consisting of $n-1$ overlapping elements of $\mathcal{L}$. The interested reader can check that the words w that arise here are the same as the "$(n-1)$-chains" of Anick [1]. There is also some similarity with the "critical $(n-1)$-stars" of Groves [11].

Suppose, for example, that M is the free commutative monoid, as in Example 1 of §2. Then the essential n-cells are the cells $(s_1, \ldots, s_n)$ with each $s_j \in S$ and $s_1 > \cdots > s_n$ in the chosen ordering on S. The corresponding words $w = s_1 \cdots s_n$ are the words in which every subword of length 2 is in $\mathcal{L}$. The reader is advised to figure out what the collapsible and redundant cells look like in this example before returning to the general theory.

We turn now to the verification of the axioms for a collapsing scheme. First, it is immediate that $c(\tau)$ is collapsible if τ is redundant. It is also quite easy to check that (C1) holds. The crux of the matter, then, is (C2). To verify this, choose a complete rewriting system R for which $\mathcal{I}$ is the set of irreducible words. Recall that this determines a relation "$\prec$" on F and that (R3) holds. Given an n-cell $\tau = (w_1, \ldots, w_n)$, let $w(\tau)$ be the word $w_1 \cdots w_n$.

LEMMA. *Let τ and τ' be redundant n-cells such that τ' is an immediate predecessor of τ. Then one of the following holds:*

(1) $w(\tau') \prec w(\tau)$.
(2) $w(\tau') = w(\tau)$, *and the maximal essential front face of τ' has higher dimension than that of τ.*

PROOF: Let $\sigma = c(\tau)$, and write $\sigma = (w_1, \ldots, w_{n+1})$. Thus

$$\tau = (w_1, \ldots, w_{i-1}, w_i w_{i+1}, w_{i+2}, \ldots, w_{n+1}),$$

where $i = i(\tau)$. [Note that $w_i w_{i+1} = w_i * w_{i+1}$ here.] By definition, then, $\tau' = d_j \sigma$ for some $j \neq i$. If $j = 0$ or $n+1$, then $w(\tau')$ is a proper subword of $w(\sigma) = w(\tau)$, and (1) holds. If $0 < j < n+1$ and $w_j w_{j+1}$ is reducible, then the computation of $d_j \sigma$ involves reducing $w_j w_{j+1}$ to $w_j * w_{j+1}$; so there is a chain of reductions from $w(\tau)$ to $w(\tau')$, and again (1) holds. Note that this case must apply if $0 < j < i$. Suppose, finally, that $i < j < n+1$ and that $w_j w_{j+1}$ is irreducible. Then $w(\tau') = w(\tau)$, and the i-dimensional front face of τ' is essential. But the maximal essential front face of τ has dimension $i-1$, so (2) holds. $\square$

It is now easy to check (**C2**). For suppose that there is a chain

$$\tau_1 \succ \tau_2 \succ \tau_3 \succ \cdots$$

of redundant n-cells. By (**R3**) we cannot have $w(\tau_j) \succ w(\tau_{j+1})$ for infinitely many j. So τ_{j+1} has a higher-dimensional essential front face than τ_j for all sufficiently large j. But this is absurd, for the dimension of the maximal essential front face of a redundant n-cell is always less than n.

Thus we have indeed constructed a collapsing scheme. Proposition 1 of §1 now yields:

THEOREM 1. *Let M be a monoid with a good set of normal forms, and let $X = |BM|$ be its classifying space. Then X admits a canonical quotient CW-complex Y, whose cells are in 1–1 correspondence with the essential cells of X. The quotient map $q : X \to Y$ is a homotopy equivalence. It maps each open essential cell of X homeomorphically onto the corresponding open cell of Y, and it maps each collapsible $(n+1)$-cell into the n-skeleton of Y.*

□

A good set $\mathcal{I}$ of normal forms will be said to have *finite type* if S is finite and $\mathcal{I}$ is the set of irreducible words associated to a complete rewriting system with only finitely many rewriting rules. It is easy to see that there are then only finitely many essential cells in each dimension. Consequently:

COROLLARY. *If M admits a good set of normal forms of finite type, then its classifying space has the homotopy type of a complex with only finitely many cells in each dimension.*

□

REMARKS. 1. In applying the results of this section to a given example, it is very likely that we will see ways to do further collapsing and thereby reduce Y to an even smaller complex. Here is one such situation that comes up fairly often: Suppose that our rewriting system contains rules $s\bar{s} \to 1$ and $\bar{s}s \to 1$ for some pair $s, \bar{s}$ of distinct generators. Assuming that s and $\bar{s}$ are irreducible, we will then have for each $n \geq 1$ a pair of essential n-cells $\sigma_n = (s, \bar{s}, s, \bar{s}, \dots)$ and $\bar{\sigma}_n = (\bar{s}, s, \bar{s}, s, \dots)$. The only non-degenerate faces of σ_n for $n \geq 2$ are $d_0\sigma_n = \bar{\sigma}_{n-1}$ and $d_n\sigma_n = \sigma_{n-1}$. We now modify our previous definitions by declaring that $\bar{\sigma}_n$ is redundant for $n \geq 1$ and that σ_n is collapsible for $n \geq 2$ (with free face $\bar{\sigma}_{n-1}$). Thus the only cell from these two infinite families that remains essential is σ_1.

2. Recall that a monoid is the same thing as a category with one object and that the classifying space construction extends to categories

(cf. [18] or [17]). For simplicity, we have confined ourselves in this section to monoids, but one can equally well treat more general categories. We will give an example of this in §7.

4. Algebraic Interpretation

In case the monoid M of §3 is a group, the cellular chain complex of the universal cover of X is the standard (or "bar") resolution of $\mathbf{Z}$ over $\mathbf{Z}M$. By taking the universal cover of the complex Y of Theorem 1, we then get a quotient complex of the bar resolution, which is a free resolution with one basis element for each essential cell of X.

We would like, more generally, to construct a "small" resolution of this type for any monoid M with a good set of normal forms, not just for groups. I do not know any way to formally deduce such a resolution from the existence of the homotopy equivalence $X \to Y$ above. What one can do, however, is simply work directly with the bar resolution and imitate the method used in §1. In other words, we do our homotopy theory in the category of chain complexes of free $\mathbf{Z}M$-modules instead of the category of CW-complexes. Here are the details.

For any monoid M, let EM be the following semi-simplicial complex: The n-simplices are $(n+1)$-tuples of elements of M, a typical such $(n+1)$-tuple being written in the form $(m_1, \ldots, m_n)m$. If $m = 1$, we suppress it from the notation and simply write $(m_1, \ldots, m_n)$. The face operators in EM are given by

$$d_i(m_1, \ldots, m_n)m = \begin{cases} (m_2, \ldots, m_n)m & i = 0 \\ (m_1, \ldots, m_i m_{i+1}, \ldots, m_n)m & 0 < i < n \\ (m_1, \ldots, m_{n-1})m_n m & i = n, \end{cases}$$

and the degeneracy operators are given by

$$s_i(m_1, \ldots, m_n)m = (m_1, \ldots, m_i, 1, m_{i+1}, \ldots, m_n)m \qquad (0 \le i \le n).$$

There is an obvious right action of M on EM by simplicial maps, where the action of m is given by $(m_1, \ldots, m_n)m' \mapsto (m_1, \ldots, m_n)m'm$. This action makes the normalized chain complex $C = C_*(EM)$ a complex of free right $\mathbf{Z}M$-modules, and, in fact, it is precisely the normalized bar resolution of $\mathbf{Z}$ over $\mathbf{Z}M$ (cf. [7], §X.2). The module C_n of n-chains has a

$\mathbf{Z}M$-basis consisting of the n-tuples $(m_1, \ldots, m_n)$ with each $m_i \neq 1$. Thus the basis elements of C correspond to the cells of $X = |BM|$. But when we compute the boundary operator $d : C_n \to C_{n-1}$ on these basis elements, we must remember to use the face operators in EM, as defined above, not those in BM. This makes a difference only for the face operator d_n.

Returning now to the situation of §3, we have a classification of the $\mathbf{Z}M$-basis elements of C as essential, collapsible, or redundant. If σ is a collapsible $(n + 1)$-cell and τ is its free face, then the boundary of σ in C can be written in the form

$$d\sigma = \pm\tau - x, \tag{1}$$

where x is a $\mathbf{Z}M$-linear combination of collapsible n-cells, essential n-cells, and redundant n-cells that are immediate predecessors of τ. [Note that σ might have degenerate faces in EM, but these are 0 in C and hence do not appear in (1).]

Suppose now that we are trying to build C by successive adjunctions, as we did for X in the proof of Theorem 1. If we proceed as in that proof, then each adjunction of a redundant basis element τ along with its associated collapsible σ yields a chain homotopy equivalence. It has a canonical homotopy inverse, which maps σ to 0 (and hence τ to $\pm x$, where x is as in (1) above).

The following analogue of Theorem 1 is now immediate:

THEOREM 2. *Let M be a monoid with a good set of normal forms, and let $C = C_*(EM)$ be its normalized bar resolution. Then C admits a canonical quotient complex D, which is a free resolution of $\mathbf{Z}$ over $\mathbf{Z}M$ with one basis element for each essential cell. The quotient map $q : C \to D$ maps each essential cell of C to the corresponding basis element of D, and it maps each collapsible cell to 0.* □

COROLLARY (ANICK–GROVES–SQUIER). *If M admits a good set of normal forms of finite type, then M is of type FP_∞, i.e., $\mathbf{Z}$ admits a free resolution D over $\mathbf{Z}M$ with D_n finitely generated for each n.* □

REMARK. The method used in this section works, with no essential change, if the ring $\mathbf{Z}M$ is replaced by an arbitrary augmented k-algebra A which comes equipped with a presentation satisfying the conditions of Bergman's diamond lemma ([2], Theorem 1.2). Here k can be any commutative ring. One starts with the normalized bar resolution C of k over A, and one obtains a quotient resolution D, with one generator for each "essential" generator of the bar resolution. In particular, we recover Anick's theorem ([1], Theorem 1.4). It seems likely that one can similarly study the homological algebra of other kinds of algebraic rewriting systems (i.e., other than monoids and associative algebras), but I have not tried to do this.

5. Computational Techniques

The statement of Theorem 2 does not contain an explicit formula for the boundary operator in D. To obtain such a formula, we need to compute the boundary $d\sigma$ of an arbitrary essential cell $\sigma \in C$ and then find the image $q(d\sigma) \in D$. The only difficulty here is that $d\sigma$ might involve some redundant cells τ. Now for each such τ, we have an element $x \in C$ such that $q(\tau) = \pm q(x)$. So the problem of computing $q(\tau)$ is reduced to the simpler problem of computing $q(x)$. This is simpler because the redundant cells that occur in x are immediate predecessors of τ. We are therefore led to the following "rewriting" procedure for explicitly describing D:

Identify D, as a graded $\mathbf{Z}M$-module, with the submodule of C generated by the essential cells. For each collapsible cell σ, write down the rule

$$\sigma \to 0.$$

For each redundant cell τ, let $\sigma = c(\tau)$, and write down the rule

$$\tau \to \pm x,$$

where x and the ambiguous sign are taken from (1). We can, of course, delete from x any terms involving collapsible cells. The remaining terms involve either essential cells or redundant cells that are immediate predecessors of τ and hence have height less than that of τ. It follows that we can use the rewriting rules above to "reduce" an element $u \in C$ to an element $\bar{u} \in D$, and this element is precisely the image $q(u)$. In particular, we can now compute the boundary operator ∂ in D; it is given by

$$\partial e = \overline{de},$$

where e is an essential cell and de is its boundary in C.

Let's make this rewriting process more explicit. Let $\tau = (w_1, \ldots, w_n)$ be redundant, and let $i = i(\tau)$. Then $\sigma = c(\tau)$ is given by

$$\sigma = (w_1, \ldots, w_{i-1}, w', w'', w_{i+1}, \ldots, w_n)$$

for some factorization $w'w''$ of w_i. The faces $d_j\sigma$ for $j > i + 1$ are either degenerate or collapsible; they can therefore be ignored. There are always at least two other faces (aside from the free face $d_i\sigma = \tau$), namely, $d_{i-1}\sigma$ and $d_{i+1}\sigma$. Set

$$\lambda = d_{i-1}\sigma = \begin{cases} (w'', w_2, \ldots, w_n) & \text{if } i = 1 \\ (w_1, \ldots, w_{i-2}, w_{i-1} * w', w'', w_{i+1}, \ldots, w_n) & \text{if } i > 1. \end{cases}$$

Roughly speaking, it is obtained from τ by pushing the left half of w_i to the left. Set

$$\rho = d_{i+1}\sigma = \begin{cases} (w_1, \ldots, w_{n-1}, w')w'' & \text{if } i = n \\ (w_1, \ldots, w_{i-1}, w', w'' * w_{i+1}, w_{i+2}, \ldots, w_n) & \text{if } i < n. \end{cases}$$

It is obtained from τ by pushing the right half of w_i to the right. The rewriting rule $\tau \to \pm x$ above now becomes

$$\tau \to \lambda + \rho + \sum_{j=0}^{i-2} (-1)^{i-j-1} d_j\sigma. \tag{2}$$

In specific examples, one often finds that the right side of (2) contains a large number of redundant cells τ', many of which are ultimately seen to satisfy $q(\tau') = 0$ after further rewriting rules are applied. If we could recognize such cells τ' in advance, then we could simply delete them from (2) and save a lot of work. With this goal in mind, we prove the following modest result:

PROPOSITION 2. Let $\tau = (w_1, \ldots, w_n)$ be an n-cell for some $n \geq 2$. If $w_1 w_2$ is irreducible, then $q(\tau) = 0$.

PROOF: We argue by induction on the length $l(w_1)$. If $l(w_1) = 1$, then τ is collapsible and there is nothing to prove. So assume that $w_1 = sw$ with $s \in S$ and $w \neq 1$. Using the notation above, we have $i = 1$, $\lambda = (w, w_2, \ldots)$, and $\rho = (s, ww_2, \ldots)$. Since ww_2 and sww_2 are irreducible, we can apply the induction hypothesis to conclude that $q(\lambda) = q(\rho) = 0$. The proposition now follows from (2). □

Even with the aid of Proposition 2, the task of computing q (and hence ∂) can be quite tedious if one simply uses (2). A better strategy is to try to guess a formula for q and then use (2) to prove inductively that the guess is correct. The induction here is with respect to the "immediate predecessor" relation, i.e., one proves the desired formula for a redundant cell τ under the assumption that the result is already known for all redundant $\tau' \prec \tau$. In other words, the result can be assumed known for all redundant cells that appear on the right side of (2). Condition **(C2)** justifies this sort of induction.

Here is a simple example. Let G be a cyclic group of finite order $m \geq 2$. It is presented (as a monoid) by the complete rewriting system with one generator s and one rule $s^m \to 1$. There is one essential n-cell $e_n = (s, s^{m-1}, s, s^{m-1}, \dots)$ in each dimension n. Moreover, one can check that the maximal essential front face of every redundant cell τ has even dimension. A straightforward induction now yields the following formula for q: Let $\tau = (w_1, \dots, w_n)$ be an n-cell. If $w_i w_{i+1}$ is irreducible for some odd $i < n$, then $q(\tau) = 0$. Otherwise $q(\tau) = e_n a$, where $a \in \mathbf{Z}G$ is given by

$$a = \begin{cases} 1 & \text{if } n \text{ is even} \\ 1 + s + \cdots + s^{k-1} & \text{if } n \text{ is odd and } w_n = s^k. \end{cases}$$

The inductive proof, which is left to the reader, is simplified by the fact that, in the notation of (2), $d_j \sigma$ is degenerate for $0 < j < i - 1$.

It now follows at once that $\partial e_n = e_{n-1} a$, where

$$a = \begin{cases} 1 - s & \text{if } n \text{ is odd} \\ 1 + s + \cdots + s^{m-1} & \text{if } n \text{ is even.} \end{cases}$$

This is the formula that one would expect (cf. [7], §XII.7, or [4], §I.6).

We close this section by deriving, for arbitrary M, the expected formula for $\partial : D_2 \to D_1$ in terms of free derivatives (cf. [4], Exercise 3 of §II.5 or Exercise 3 of §IV.2). To simplify the statement, we assume that **(R4)** holds. Let F, as before, be the free monoid generated by S. There is a unique function $\delta : F \to D_1$ such that $\delta(s) = (s)$ for all $s \in S$ and $\delta(uv) = \delta(u)r(v) + \delta(v)$ for all $u, v \in F$. For any 1-cell $\tau = (w)$, I claim that $q(\tau) = \delta(w) \in D_1$. To prove this, we may assume that $w = sw'$ with $s \in S$ and $w' \neq 1$, and we may assume inductively that $q(w') = \delta(w')$.

Then

$$q(\tau) = q(w') + (s)w' \qquad \text{by (2)}$$
$$= \delta(w') + \delta(s)w'$$
$$= \delta(w),$$

as claimed.

Recall that, since we have assumed **(R4)**, there is one essential 2-cell for each $(w_1, w_2) \in R$, where R is the canonical rewriting system for $\mathcal{I}$.

PROPOSITION 3. *Let e be the essential 2-cell corresponding to a rewriting rule $(w_1, w_2) \in R$. Then $\partial e = \delta(w_1) - \delta(w_2)$.*

PROOF: We have $e = (s, w)$, with $w_1 = sw$ and $w_2 = r(sw) = s * w$. Hence $\partial e = q(de) = q(w) - q(w_2) + (s)w$. Now $q(w_2) = \delta(w_2)$, and $q(w) + (s)w = \delta(sw) = \delta(w_1)$, whence the proposition. □

6. The Case of Conjugation Relations

This section is motivated by [6] and by some unpublished work of Craig Squier, in which "cubical" resolutions were constructed for a number of monoids.

We continue to assume that M is a monoid with a good set $\mathcal{I}$ of normal forms. Suppose that the generating set S comes equipped with a total order and that the elements of $\mathcal{I}$ are the non-decreasing words $s_1 \cdots s_n$, as in the case of the free commutative monoid (Example 1 of §2). The canonical rewriting system R for $\mathcal{I}$ then has one rule $ts \to r(ts)$ for each pair of generators t, s with $t > s$. Suppose further that s occurs as the first letter in $r(ts)$, so that the rule has the form

$$ts \to su,$$

where u is an irreducible word whose first letter is greater than or equal to s. Following the customary notation for conjugates in group theory, we will write $u = t^s$. In the case of the free commutative monoid, for instance, we have $t^s = t$. The reader may wish to concentrate on this case, at least on first reading.

For simplicity, we will assume that $l(t^s) = 1$, i.e., that $t^s \in S$. Some of what we do is valid more generally, but the statements become more complicated. We will also assume that the "conjugation function" satisfies

$$s < t_1 < t_2 \implies t_1^s < t_2^s.$$

It is convenient to extend the conjugation notation slightly: Given $s \in S$ and a word $w = t_1 \cdots t_n$ such that $t_i > s$ for all i, set

$$w^s = t_1^s \cdots t_n^s.$$

Note, then, that $r(ws) = sw^s$ if w is irreducible. Next, given $t \in S$ we wish to define $t^w \in S$ for certain words $w = s_1 \cdots s_n$. Proceeding by induction on n, let $u = s_1 \cdots s_{n-1}$. We then set

$$t^w = (t^u)^{s_n},$$

provided t^u is defined and $t^u > s_n$.

We now apply the method of §§4 and 5 to M. The essential cells of $X = |BM|$ are the cells $(s_1, \ldots, s_n)$ with $s_i \in S$ and $s_1 > \cdots > s_n$. We therefore obtain a free resolution D with one generator for each such cell. In order to describe the boundary operator ∂ in D, we introduce some operators A_i and B_i which map D_n to D_{n-1}. They are to be $\mathbf{Z}M$-linear, so we need only specify them on the essential n-cells. Given an essential cell $\sigma = (s_1, \ldots, s_n)$ and an integer i with $1 \le i \le n$, set

$$A_i \sigma = (s_1^{s_i}, \ldots, s_{i-1}^{s_i}, s_{i+1}, \ldots, s_n)$$
$$B_i \sigma = (s_1, \ldots, s_{i-1}, s_{i+1}, \ldots, s_n) s_i^u,$$

where $u = s_{i+1} \cdots s_n$. (Note that, in view of our assumptions, the right-hand sides are well-defined and in D_{n-1}.) In terms of the word $s_1 \cdots s_n$ associated to σ, we compute $A_i \sigma$ by moving s_i to the left and then deleting it, and we compute $B_i \sigma$ by moving s_i to the right and then retaining the resulting conjugate of s_i as an operator.

PROPOSITION 4. *The boundary operator* $\partial : D_n \to D_{n-1}$ *is given by*

$$\partial = \sum_{i=1}^{n} (-1)^{i-1} (A_i - B_i).$$

To prove this, it is convenient to extend the operators A_i and B_i to arbitrary cells $\sigma = (w_1, \ldots, w_n)$. The definitions are slightly more complicated, because it might not be possible to move w_i all the way to the left and right. What we do, roughly speaking, is move them as far as possible. The precise definitions follow.

We define $A_i\sigma$ for $1 \le i \le n$ by induction on i. Consider the face

$$d_{i-1}\sigma = \begin{cases} (w_2, \ldots, w_n) & \text{if } i = 1 \\ (w_1, \ldots, w_{i-1} * w_i, \ldots, w_n) & \text{if } i > 1. \end{cases}$$

We set $A_i\sigma = d_{i-1}\sigma$ if any of the following conditions hold: (a) $i = 1$; (b) $l(w_i) > 1$; (c) $i > 1$, $l(w_i) = 1$, and there is a letter t occurring in w_{i-1} such $t \le w_i$ in the given ordering on S. If (a)–(c) all fail, then $i > 1$ and

$$\sigma = (w_1, \ldots, w_{i-1}, s, w_{i+1}, \ldots, w_n)$$

with $s \in S$ and $t > s$ for all t occurring in w_{i-1}. Hence

$$d_{i-1}\sigma = (w_1, \ldots, w_{i-2}, sw_{i-1}^s, w_{i+1}, \ldots, w_n).$$

We now set $A_i\sigma = A_{i-1}\sigma'$, where

$$\sigma' = (w_1, \ldots, w_{i-2}, s, w_{i-1}^s, w_{i+1}, \ldots, w_n).$$

Similarly, B_i is defined for $1 \le i \le n$ by descending induction on i. Consider the face

$$d_i\sigma = \begin{cases} (w_1, \ldots, w_{n-1})w_n & \text{if } i = n \\ (w_1, \ldots, w_i * w_{i+1}, \ldots, w_n) & \text{if } i < n. \end{cases}$$

We set $B_i\sigma = d_i\sigma$ if any of the following conditions hold: (a) $i = n$; (b) $l(w_{i+1}) > 1$; (c) $i < n$, $l(w_{i+1}) = 1$, and $t \le w_{i+1}$ for some t occurring in w_i. If (a)–(c) all fail, then $i < n$ and

$$\sigma = (w_1, \ldots, w_i, s, w_{i+2}, \ldots, w_n)$$

with $s \in S$ and $t > s$ for all t occurring in w_i. We now have

$$d_i\sigma = (w_1, \ldots, w_{i-1}, sw_i^s, w_{i+2}, \ldots, w_n),$$

and we set $B_i\sigma = B_{i+1}\sigma'$, where

$$\sigma' = (w_1, \ldots, w_{i-1}, s, w_i^s, w_{i+2}, \ldots, w_n).$$

The proof of the proposition is based on the following lemma:

LEMMA. *Let τ be a redundant n-cell, let $i = i(\tau)$, and let $\sigma = c(\tau)$. Then $q(\tau) = q(A_i\sigma + B_{i+1}\sigma)$.*

This shows that the rewriting rule (2) of §5 can be replaced by the much simpler rule

$$\tau \to A_i\sigma + B_{i+1}\sigma.$$

Accepting the lemma for the moment, we can easily prove the proposition. For suppose $e = (s_1, \ldots, s_n)$ is essential, and consider $d_j e$ for $0 < j < n$. We have

$$d_j e = (s_1, \ldots, s_j * s_{j+1}, \ldots, s_n)$$
$$= (s_1, \ldots, s_{j-1}, s_{j+1}s_j^{s_{j+1}}, s_{j+2}, \ldots, s_n),$$

and the lemma yields $q(d_j e) = A_{j+1}e + B_j e$. Since $d_0 e = A_1 e$ and $d_n e = B_n e$, we conclude that

$$\partial e = A_1 e + (-1)^n B_n e + \sum_{j=1}^{n-1}(-1)^j q(d_j e)$$

$$= A_1 e + (-1)^n B_n e + \sum_{j=1}^{n-1}(-1)^j (A_{j+1}e + B_j e)$$

$$= \sum_{j=1}^{n}(-1)^{j-1}(A_j e - B_j e),$$

as required.

PROOF OF THE LEMMA: We argue by induction with respect to the "immediate predecessor" relation, as in §5. Thus we may assume that the lemma is already known for all redundant τ' that appear on the right side of (2). Using this induction hypothesis, one can check that, in the notation of (2), $q(\rho) = q(B_{i+1}\sigma)$, $q(\lambda) = q(A_i\sigma)$, and $q(d_j\sigma) = 0$ for $j < i - 1$, whence the lemma. Here, for instance, is the proof of the assertion about $\rho = d_{i+1}\sigma$; the proofs of the other two assertions are similar and are left to the reader.

Write $\tau = (s_1, \ldots, s_{i-1}, s_i w, w_{i+1}, \ldots, w_n)$, where $s_1 > \cdots > s_i$ in S and w is a non-trivial irreducible word whose first letter is greater than or equal to s_i. Then $\sigma = (s_1, \ldots, s_i, w, w_{i+1}, \ldots, w_n)$. The result to be proved is trivial in the cases where $B_{i+1}\sigma$ was defined to be $d_{i+1}\sigma$. So we may assume that $i < n$ and that $\sigma = (s_1, \ldots, s_i, w, s, \ldots)$, with $s \in S$ and $t > s$ for all t which occur in w. Then $\rho = (s_1, \ldots, s_i, sw^s, \ldots)$. This is collapsible if $s_i \leq s$, in which case $q(\rho) = 0 = q(B_{i+1}\sigma)$. If $s_i > s$, on the other hand, then ρ is redundant, with $i(\rho) = i + 1$. Let

$$\sigma' = c(\rho) = (s_1, \ldots, s_i, s, w^s, \ldots).$$

We have $A_{i+1}\sigma' = (s_1^s, \ldots, s_i^s, w^s, \ldots)$, which is collapsible. So the induction hypothesis yields

$$\begin{aligned} q(\rho) &= q(A_{i+1}\sigma' + B_{i+2}\sigma') \\ &= q(B_{i+2}\sigma') \\ &= q(B_{i+1}\sigma), \end{aligned}$$

as required.

7. Examples

We have already treated the case of a finite cyclic group in §5. We present here a few more examples to illustrate how the methods of this paper can be used to obtain explicit Eilenberg–MacLane complexes, or explicit free resolutions, for various groups or monoids. When we are interested in a group G, however, we will often get results about G by studying a suitable submonoid $M \subset G$. (This idea is suggested by the work of Craig Squier.) And in one example, instead of replacing G by a submonoid, we will replace G by a category $\mathcal{M}$ and apply our methods to the classifying space $|B\mathcal{M}|$.

The examples are only intended to illustrate the method; in all cases except possibly the last, the results obtained were already known.

Example 1: Free groups. Let F be the free group generated by a set S. The usual normal forms for the elements of F are obtained from a monoid rewriting system with generators $s, \bar{s}$ ($s \in S$) and rules $s\bar{s} \to 1$ and $\bar{s}s \to 1$. The only essential cells in positive dimensions are those of the form $(s, \bar{s}, s, \bar{s}, \ldots)$ or $(\bar{s}, s, \bar{s}, s, \ldots)$, which we discussed in Remark 1 at the end of §3. As we explained there, it is possible to modify the essential/collapsible/redundant classification in order to "collapse away" most of these cells. The result, then, is the usual $K(F, 1)$-complex Y, with one 1-cell for each $s \in S$ and no higher-dimensional cells.

Example 2: Free abelian groups. Let G be the free abelian group generated by a set S. The easiest way to proceed here is to work not with G directly, but rather with the submonoid M generated by S. Note that $G =$

MM^{-1}, i.e., every element of G can be written as $m_1 m_2^{-1}$ with $m_i \in M$. It is not hard to show, in this situation, that $|BM|$ is homotopy equivalent to $|BG|$ (cf. [10], Proposition 4.4). So there is no harm in replacing G by M. From the point of view of homological algebra, the content of this statement is that $\mathbf{Z}G$ is flat as a left $\mathbf{Z}M$-module (cf. [7], Chapter X, proof of Proposition 4.1), so if we have a resolution D of $\mathbf{Z}$ by free right $\mathbf{Z}M$-modules, then $D \otimes_{\mathbf{Z}M} \mathbf{Z}G$ is a resolution of $\mathbf{Z}$ by free right $\mathbf{Z}G$-modules.

Choose a total order on S, and use the normal forms of Example 1 of §2. We have already noted that this fits into the framework of §6, so we obtain a free resolution D of $\mathbf{Z}$ over $\mathbf{Z}M$ with one basis element for every decreasing sequence $s_1 > \cdots > s_n$ of elements of S. The boundary operator $\partial : D_n \to D_{n-1}$ is given by

$$\partial = \sum_{i=1}^{n} (-1)^{i-1}(A_i - B_i),$$

where

$$A_i(s_1, \ldots, s_n) = (s_1, \ldots, s_{i-1}, s_{i+1}, \ldots, s_n)$$
$$B_i(s_1, \ldots, s_n) = (s_1, \ldots, s_{i-1}, s_{i+1}, \ldots, s_n)s_i.$$

This resolution is well-known. See, for instance, [7], §X.5.

Example 3: A one-relator group. Let G be the group with two generators a, t and the single relation $a^t = a^2$, where $a^t = t^{-1}at$. Let M be the submonoid generated by a and t. As in Example 2, we have $G = MM^{-1}$, so we can obtain a $K(G,1)$ by working with M instead of G. We have already given a good set of normal forms for M (Example 2 of §2). One checks that the only essential cells are $(\ )$, (a), (t), and (a,t), so Theorem 1 yields a $K(G,1)$ with one vertex, two 1-cells, and one 2-cell. The corresponding resolution $D \otimes_{\mathbf{Z}M} \mathbf{Z}G$ is the famous Lyndon resolution [15].

Example 4: Thompson's group. Let G be the group with infinitely many generators $x_0, x_1, x_2, \ldots$ and infinitely many relations $x_j^{x_i} = x_{j+1}$ for $i < j$. This group was first introduced by R. Thompson, and it was the main object of study in [6], where it was called F. See [6] and [5] for further references and a discussion of the history of this group. As in Examples 2 and 3, we will apply our methods to the submonoid $M \subset G$ generated by the x_i. Once again, we have $G = MM^{-1}$, so there is no harm in doing this.

The elements of M have normal forms $x_{i_1} \cdots x_{i_n}$ with $i_1 \leq \cdots \leq i_n$ (cf. [6], 1.3), and these normal forms arise from the complete rewriting system with rules $x_j x_i \rightarrow x_i x_{j+1}$ for $i < j$. As in Example 2, then, we are in the situation of §6. To simplify the notation, let's write $(i_1, \ldots, i_n)$ for the generator of D corresponding to the essential cell $(x_{i_1}, \ldots, x_{i_n})$. The result, then, is that D has a basis consisting of decreasing sequences $i_1 > \cdots > i_n$ of non-negative integers, and $\partial : D_n \rightarrow D_{n-1}$ is given by

$$\partial = \sum_{j=1}^{n} (-1)^{j-1}(A_j - B_j),$$

where

$$A_j(i_1, \ldots, i_n) = (i_1 + 1, \ldots, i_{j-1} + 1, i_{j+1}, \ldots, i_n)$$
$$B_j(i_1, \ldots, i_n) = (i_1, \ldots, i_{j-1}, i_{j+1}, \ldots, i_n)x_{i_j + n - j}.$$

Note that the irreducible words $r(w(\sigma))$ corresponding to the essential cells σ are precisely the words $x_{q_1} \cdots x_{q_n}$ with $q_{j+1} \geq q_j + 2$ for $j = 1, \ldots, n - 1$. If we rewrite the A and B operators in terms of the sequences $(q_1, \ldots, q_n)$, we see that our resolution $D \otimes_{\mathbf{Z}M} \mathbf{Z}G$ is isomorphic to the "big" resolution constructed in [6]. In particular, we have obtained a new proof of Theorem 4.1 of [6]. Geoghegan and I went on in [6] to collapse this big resolution to one with only two generators in each positive dimension. This further collapse was based on the combinatorics of the "cubical face operators" A_j and B_j; I do not know how to explain it in terms of rewriting systems.

Example 5: Thompson's group again. Let G be the group of piecewise linear homeomorphisms g of the unit interval $[0, 1]$ with the following two properties: (a) the singularities of g occur at dyadic rational numbers, i.e., at points in $\mathbf{Z}[1/2]$; and (b) for any non-singular x, one has $g'(x) = 2^n$ for some $n \in \mathbf{Z}$. It is known that G is isomorphic to the group of Example 4 (see, for instance, [5], Propositions 4.1, 4.4, and 4.8). Our approach this time, however, will be to construct, directly from the definition of G as a homeomorphism group, a category $\mathcal{M}$ whose classifying space is a $K(G, 1)$-complex. We will then apply the methods of §3 to $\mathcal{M}$ (cf. Remark 2 at the end of §3). Our discussion will be sketchy and will assume familiarity with the definition and basic properties of the classifying space of a category, as given in [18] or [17].

Let $\mathcal{G}$ be the groupoid whose objects are the intervals I of the form $[0, l]$, where l is a positive integer, and whose morphisms $I \rightarrow I'$ are the piecewise linear homeomorphisms satisfying conditions (a) and (b) above. Then $\mathcal{G}$ is connected (i.e., all objects are isomorphic), and our group G is the group of maps from $[0,1]$ to itself in $\mathcal{G}$. Hence the classifying space of $\mathcal{G}$ is a $K(G,1)$-complex.

By an *admissible subdivision* of $[0,l]$ we will mean a subdivision obtained by starting with the standard partition into l subintervals $[i-1,i]$ $(1 \leq i \leq l)$ and then repeating 0 or more times the operation of inserting a midpoint into a subinterval. One can show that every map $g : I \to I'$ in $\mathcal{G}$ can be described in terms of admissible subdivisions, i.e., there are admissible subdivisions of I and I' (into the same number of subintervals) such that g maps the intervals of the first subdivision linearly to the intervals of the second (cf. [5], proof of Proposition 4.4).

Call g *positive* if it admits a description of this form in which the subdivision of I' is the standard one with l' subintervals, where $I' = [0,l']$. Note, then, that a general map $g : I \to I'$ in $\mathcal{G}$ is a composite $p^{-1}q$, where $q : I \to I''$ and $p : I' \to I''$ are positive maps with the same target; indeed, this is just a restatement of the fact that g is describable in terms of admissible subdivisions of I and I'.

Identity maps are positive, and a composite of positive maps is positive; so $\mathcal{G}$ has a subcategory $\mathcal{M}$ whose maps are the positive maps. We can now express the result of the previous paragraph by writing $\mathcal{G} = \mathcal{M}^{-1}\mathcal{M}$. This implies that $|B\mathcal{M}|$ is a $K(G,1)$-complex.

[Sketch of proof: Let $\mathcal{E}$ be the following category: An object of $\mathcal{E}$ is a map $g : [0,1] \to I$ in $\mathcal{G}$, where I is an arbitrary object of $\mathcal{G}$; given $g : [0,1] \to I$ and $g' : [0,1] \to I'$, a map from g to g' in $\mathcal{E}$ is a map $m : I \to I'$ in $\mathcal{M}$ such that $g' = mg$. Using the equation $\mathcal{G} = \mathcal{M}^{-1}\mathcal{M}$, one shows that for any two objects g, g' of $\mathcal{E}$ there is a third object g'' to which they both map. It follows that $\mathcal{E}$ is a filtering category and hence that $|B\mathcal{E}|$ is contractible. Our assertion that $|B\mathcal{M}|$ is a $K(G,1)$-complex is now immediate, since G acts freely on $|B\mathcal{E}|$ with $|B\mathcal{M}|$ as quotient. Alternatively, one can deduce the assertion from Quillen's Theorem A, applied to the inclusion $\mathcal{M} \hookrightarrow \mathcal{G}$.]

For any $l \geq 1$ and any i with $1 \leq i \leq l$, let $\delta_i^l : [0,l] \to [0,l+1]$ be the map in $\mathcal{M}$ which has slope 2 on $[i-1,i]$ and slope 1 elsewhere. To simplify the notation, we will often indicate this map by writing $[0,l] \xrightarrow{i} [0,l+1]$. The maps δ_i^l generate $\mathcal{M}$. In fact, every map $[0,l] \to [0,l+n]$ in $\mathcal{M}$ is uniquely expressible as a composite

$$[0,l] \xrightarrow{i_1} [0,l+1] \xrightarrow{i_2} \cdots \xrightarrow{i_n} [0,l+n]$$

with $i_1 \leq \cdots \leq i_n$. Moreover, these normal forms for the maps in $\mathcal{M}$ are produced by the rewriting rules

$$\delta_i^{l+1}\delta_j^l \rightarrow \delta_{j+1}^{l+1}\delta_i^l \qquad (1 \leq i < j \leq l).$$

Recall now that the classifying space X of $\mathcal{M}$ is the geometric realization of a semi-simplicial complex $B\mathcal{M}$ with one n-simplex for every diagram $I_0 \rightarrow \cdots \rightarrow I_n$ consisting of n composable maps. (When $n = 0$, such a "diagram" is simply an object of $\mathcal{M}$.) Using the methods of §3, one constructs a collapsing scheme for $B\mathcal{M}$ whose essential cells are those of the form

$$[0, l] \xrightarrow{i_1} \cdots \xrightarrow{i_n} [0, l+n]$$

with $i_1 > \cdots > i_n$. Thus we can collapse X to a $K(G, 1)$-complex Y with one n-cell for every sequence $l \geq i_1 > \cdots > i_n$ of positive integers. The corresponding resolution D of $\mathbf{Z}$ over $\mathbf{Z}G$ is a quotient of $C = C_*(B\mathcal{E})$, where $\mathcal{E}$ is the category introduced above.

REMARKS. 1. This $K(G, 1)$-complex Y is essentially the same as one that was obtained by Melanie Stein [unpublished], using different methods. As in Example 4, it is easy to collapse Y further, to a complex with one vertex and exactly two n-cells for each $n \geq 1$. The complex Y and analogous complexes for other homeomorphism groups have proved to be quite useful, for reasons that will be explained elsewhere.

2. The category $\mathcal{E}$ which arose above is actually a poset viewed as a category. It is isomorphic to the poset of ordered bases that was used in [5].

Example 6: A group of Brin and Squier. Our last example is intended to illustrate the advantage of using the "right" set of generators, or the "right" submonoid, for a group G. Let G be the group of piecewise linear homeomorphisms g of the half-line $[0, \infty)$ such that g has only finitely many singularities and satisfies the slope and singularity conditions (a) and (b) of Example 5. This is the group called $G(2)$ in [3]. It admits a presentation with infinitely many generators $x_0, x_1, x_2, \ldots$ and infinitely many relations $x_j^{x_i} = x_{2j-i}$ for $i < j$, cf. [3], (2.11). We can now proceed exactly as in Example 4 to obtain a "cubical" free resolution D with one generator for each decreasing sequence $i_1 > \cdots > i_n$ of non-negative integers.

Now it is known that G is of type FP_∞; in fact, G is an ascending HNN extension of Thompson's group (Example 4). So one might expect to be able to use the cubical face operators A_i and B_i to collapse D to a small resolution, as in [6]. But I have not been able to do this. The situation changes drastically, however, if we replace the generators x_i by new generators $t, y_0, y_1, \ldots$, where $t = x_0$ and $y_i = x_{i+1}^{-1} x_i = x_i x_{i+2}^{-1}$. With these new generators, the defining relations become

$$y_i^t = y_{2i} y_{2i+2}$$
$$y_j^{y_i} = y_{j+1} \qquad \text{for } i < j.$$

This presentation reflects the fact that G is an HNN extension of Thompson's group.

It is not hard to show that the submonoid M of G generated by t and the y_i admits a presentation by a complete rewriting system with rules $y_i t \to t y_{2i} y_{2i+2}$ and $y_j y_i \to y_i y_{j+1}$ for $i < j$. And, again, we have $G = MM^{-1}$. So we obtain a resolution with basis elements $(y_{i_1}, \ldots, y_{i_n})$ and $(y_{i_1}, \ldots, y_{i_{n-1}}, t)$, where the subscripts are strictly decreasing. Our rewriting system does not quite fit the framework of §6, because of the presence of two y's on the right side of the first rule. But one can still use the methods of §6 to obtain a cubical boundary formula $\partial = \sum_{i=1}^{n} (-1)^{i-1} (A_i - B_i)$, where the A_i and B_i are defined as before by moving generators to the right or left. The only difference is that if σ is an essential cell of the form $(y_{i_1}, \ldots, y_{i_{n-1}}, t)$, then the result of moving t to the left is $\tau = (y_{2i_1} y_{2i_1+2}, \ldots, y_{2i_{n-1}} y_{2i_{n-1}+2})$, which is not essential. Thus $A_n \sigma$ has to be interpreted as $q(\tau)$ rather than τ. Since t does not occur in τ, we can compute $q(\tau)$ by repeated applications of the lemma in §6. This will yield a linear combination of essential cells involving y's only, and not t. It is not necessary, for our present purposes, to know any more about $q(\tau)$.

It is now a simple matter to imitate [6] and collapse the resolution D to one with only finitely many generators in each dimension. Here are some hints as to how to do this: As essential cells in dimensions 1 and 2, take (t), (y_0), (y_1), (y_2, y_0), (y_3, y_1), (y_0, t), and (y_1, t). To collapse the 1-cell $\tau = (y_i)$, which is now considered redundant if $i \geq 2$, set $c(\tau) = (y_{i-1}, y_{i-2})$; note that $A_2 c(\tau) = \tau$. Similarly, collapse the 2-cell $\tau = (y_i, t)$ with $i \geq 2$ by setting $c(\tau) = (y_{i-1}, y_{i-2}, t)$. Suppose, finally, that $\tau = (y_j, y_i)$ with either (a) $j \geq i + 3$ or (b) $j = i + 2$ and $i \geq 2$. We then take $c(\tau)$ to be (y_{j-1}, y_{j-2}, y_i) in case (a) and $(y_{j-1}, y_{i-1}, y_{i-2})$ in case (b). Further details are omitted.

REFERENCES

1. D. J. Anick, *On the homology of associative algebras*, Trans. Amer. Math. Soc. **296** (1986), 641–659.
2. G. M. Bergman, *The diamond lemma for ring theory*, Advances in Math. **29** (1978), 178–218.
3. M. G. Brin and C. C. Squier, *Groups of piecewise linear homeomorphisms of the real line*, Invent. Math. **79** (1985), 485–498.
4. K. S. Brown, "Cohomology of Groups," Graduate Texts in Mathematics 87, Springer-Verlag, New York, 1982.
5. K. S. Brown, *Finiteness properties of groups*, J. Pure Appl. Algebra **44** (1987), 45–75.
6. K. S. Brown and R. Geoghegan, *An infinite-dimensional torsion-free FP_∞ group*, Invent. Math. **77** (1984), 367–381.
7. H. Cartan and S. Eilenberg, "Homological Algebra," Princeton University Press, Princeton, 1956.
8. B. Eckmann, *Der Cohomologie-Ring einer beliebigen Gruppe*, Comment. Math. Helv. **18** (1946), 232–282.
9. S. Eilenberg and S. MacLane, *Relations between homology and homotopy groups of spaces*, Ann. of Math. (2) **46** (1945), 480–509.
10. Z. Fiedorowicz, *Classifying spaces of topological monoids and categories*, Amer. J. Math. **106** (1984), 301–350.
11. J. R. J. Groves, *Rewriting systems and resolutions over group rings*, in preparation.
12. J. R. J. Groves and G. C Smith, *Rewriting systems and soluble groups*, preprint.
13. D. E. Knuth, "The Art of Computer Programming," Volume 1, Fundamental algorithms, second edition, Addison–Wesley, Reading, 1973.
14. P. Le Chenadec, "Canonical Forms in Finitely Presented Algebras," Research Notes in Theoretical Computer Science, Pitman (London-Boston) and Wiley (New York), 1986.
15. R. C. Lyndon, *Cohomology theory of groups with a single defining relation*, Ann. of Math. **52** (1950), 650–665.
16. D. McDuff, *On the classifying spaces of discrete monoids*, Topology **18** (1979), 313–320.
17. D. Quillen, *Higher algebraic K-theory. I*, in "Algebraic K-theory, I: Higher K-theories," Proc. Conf., Battelle Memorial Inst., Seattle, 1972, Lecture Notes in Math., Vol. 341, Springer, Berlin, 1973, pp. 85–147.
18. G. B. Segal, *Classifying spaces and spectral sequences*, Inst. Hautes Études Sci. Publ. Math. **34** (1968), 105–112.
19. C. C. Squier, *Word problems and a homological finiteness condition for monoids*, J. Pure Appl. Algebra **49** (1987), 201–217.

Mathematics Department, White Hall, Cornell University, Ithaca, NY 14853

Research at MSRI partially supported by a grant from the National Science Foundation.

Combings of Groups

Abstract. We use combings to prove a theorem à la Rips. Applying it to
hyperbolic groups one recovers Rips' Theorem. When applied to automatic
groups the theorem yields, via a theorem of Ken Brown, a proof that automatic groups are finitely presented and of type FP_∞. We also characterize
hyperbolic groups in terms of combings, and discuss the relationship between the combings considered here and isoperimetric inequalities.

Recall that for hyperbolic groups (in the sense of Gromov, see [Gr],
[BGHHSST], [CDP], [S]) one has the following

THEOREM [RIPS]. *Let G be a δ-hyperbolic group. Then for $d \geq 4\delta + 1$,
there is a simplicial complex $P_d(G)$ with the following properties:*

(1) $P_d(G)$ *is contractible, locally finite and finite dimensional.*
(2) *The simplicial G-action on $P_d(G)$ is free and transitive on the set of
vertices.*
(3) $P_d(G)/G$ *is compact.*

Recall that a group G is said to be of type FP_∞ if the $\mathbb{Z}G$-module $\mathbb{Z}$
admits a projective resolution which is finitely generated in all dimensions.
Rips' Theorem implies, in particular, that hyperbolic groups are finitely
presented and of type FP_∞, and that torsion-free hyperbolic groups have
finite cohomological dimension.

Thurston has abstracted the notion of *combing* from properties of automatic groups (see [CEHPT]). We define several types of combings below.
In fact, as a byproduct of our study we obtain yet another definition of
hyperbolic groups, in terms of combings (see Theorem 4 below).

We adapt Rips' idea to the case of groups that admit a *bounded combing*
(see §1) and prove a result (Theorem 1 below) that can be applied both
to hyperbolic groups (in which case one recovers Rips' Theorem) and to
automatic groups. In this last case one obtains a condition that Ken Brown
(see [B]) has shown implies that the group is finitely presented and of type
FP_∞. This is an unpublished result of Thurston.

Theorem 1 leaves open the question as to whether torsion-free automatic
groups must have finite cohomological dimension. The first example of

---

*Supported by a grant of the Swedish Natural Science Research Council (NFR).

a finitely presented, infinite dimensional torsion-free group of type FP_∞ was given by Brown and Geoghegan in [BG]. Is this group automatic? S. Gersten [Ge2] has indicated that this question is open. Several other infinite families of groups of this type are studied in [B], and whether or not these are automatic is also unknown.

This note grew out of an exposition of Rips' Theorem that I recently gave at M.S.R.I.'s seminar on negatively curved (or hyperbolic) groups. I am grateful for the remarks made by several people attending the seminar.

1. Combings

Let G be a group and $A = A^{-1}$ a finite set of semigroup generators. We use A to make G into a metric space by setting $d(g, h) = \ell_A(g^{-1}h)$, where ℓ_A is the word length in G relative to A.

Given G and A as above, define a *path* in G from 1 to g to be a function $p : [0, n_g] \to G$ defined on some initial segment of $\mathbf{N}$, satisfying $p(0) = 1$, $p(n_g) = g$ and $d(p(t), p(t+1)) \le 1$ for all $t \in [0, n_g - 1]$. We extend p to a function $\mathbf{N} \to G$ by setting $p(t) = p(n_g)$ for all $t \ge n_g$. We always assume that n_g is the smallest number with this property. Let $\mathcal{P}(G)$ denote the set of all paths in G based at 1, and let $\pi : \mathcal{P}(G) \to G$ be the evaluation map $\pi(p) = p(n_g) = g$. We let $\mathcal{G}(G) \subset \mathcal{P}(G)$ denote the set of all *geodesic paths* in G based at 1. Thus if $p : \mathbf{N} \to G$ is geodesic, then $d(p(t_1), p(t_2)) = |t_1 - t_2|$, for all $t_1, t_2 \in [0, n_g]$. In particular, $n_g = d(g, 1)$.

A *combing* of G (with respect to A) is any section $s : G \to \mathcal{P}(G)$ of π. A combing s is called *geodesic* if im $s \subset \mathcal{G}(G)$. Note that G always has a geodesic combing. Let $\mathbf{N}_+ = \{1, 2, \dots\}$. A *bounded combing* is a combing s satisfying the following condition: there is a monotone function $\varphi : \mathbf{N} \to \mathbf{N}_+$ with $\varphi(n) \ge n$, such that for all $g, h \in G$ and $t \in \mathbf{N}$

$$(1) \qquad d\left(s(g)(t), s(h)(t)\right) \le \varphi\left(d(g, h)\right)$$

Automatic groups admit bounded combings. In fact by Theorem 7.2 of [CEHPT], they admit a combing s for which there are constants C_1, C_2 such that

$$d(s(g)(t), s(h)(t)) \le C_1 d(g, h) + C_2$$

for all $t \in \mathbf{N}$. These combings are called *quasi-Lipschitz* by Thurston.

For our purposes it will be convenient to use the following equivalent definition of bounded combing (see Lemma 1 below):

$\exists$ a monote function $\varphi : \mathsf{N} \to \mathsf{N}_+$ with $\varphi(n) \geq n$, and $C \in \mathsf{N}, C \geq 2$

such that

$$\forall g, h \in G, \ \forall t, t' \in \mathsf{N} \text{ if } 0 \leq t - t' \leq [C/2] \leq t, \text{ then}$$

(1')
$$d(s(g)(t - [C/2]), s(h)(t')) \leq \varphi(d(g, h))$$

where $[C/2]$ denotes the integral part of $C/2$.

LEMMA 1. Let s be a combing of G. Then the following conditions are equivalent:

   (i) s is bounded.
   (ii) s satisfies (1').
   (iii) s is quasi-Lipschitz.

PROOF: $(i) \implies (ii)$. Let C be any integer ≥ 2, and suppose that t, t' satisfy $0 \leq t - t' \leq [C/2] \leq t$. Then for any $g, h \in G$,

$$d(s(g)(t - [C/2]), s(h)(t')) \leq d(s(g)(t - [C/2]), s(g)(t))$$
$$+ d(s(g)(t), s(h)(t)) + d(s(h)(t), s(h)(t'))$$
$$\leq [C/2] + \varphi(d(g, h)) + t - t' \leq \varphi(d(g, h)) + C$$

so that (1') holds (for $\bar{\varphi} = \varphi + C$). The proof of the converse is similar.

To see that $(i) \implies (iii)$, suppose that $d(g, h) = n$ and choose a geodesic path $g = g_0, g_1, \ldots, g_n = h$. Then $d(s(g_i)(t), s(g_{i+1})(t)) \leq \varphi(d(g_i, g_{i+1})) = \varphi(1)$. Hence $d(s(g)(t), s(h)(t)) \leq n\varphi(1) = \varphi(1)d(g, h)$ as desired. The converse is obvious.

A combing s will be called *contracting* if there is a constant $C \in \mathsf{N}$, $C \geq 2$ such that for all $g, h \in G$, $t, t' \in \mathsf{N}$, if $t' \leq t$, $t \in [[C/2], n_g]$ and $d(s(g)(t), s(h)(t')) \leq C$ then

(2)
$$d(s(g)(t - [C/2]), s(h)(t')) \leq C$$

Finally, we say that a combing s is *C-r-contracting* if it is geodesic and there are constants $C \in \mathsf{N}, C \geq 2$ and $r \in (1/2, 1) \subset \mathsf{R}$ such that for all $g, h \in G$ and $t, t', k \in \mathbf{N}$, if $t' \leq t$, $t \in [[C/2], n_g]$, $k \in [2, C]$, and $d(s(g)(t), s(h)(t')) \leq k$ then

(3) $\qquad d(s(g)(t - [k/2]), s(h)(t')) \leq (r - 1/2)C + [k/2]$

We show in §3 that G is hyperbolic if and only if it admits a C-r-contracting combing for some r, $1/2 < r \leq 3/5$. The relationship between these notions is

$$C\text{-}r\text{-contracting} \implies \text{contracting} \implies \text{bounded}$$

The first implication is obvious, but I don't know if the converse is true. For the second implication we have

LEMMA 2. *Every contracting combing is bounded.*

PROOF: Assuming (2), we will prove that (1') holds for the same constant C and with $\varphi(0) = 3C$, and $\varphi(n) = 3(m+1)C$ whenever $mC < n \leq (m+1)C$ $(m \in \mathbf{N})$. Given $g, h \in G$ there is $n \in \mathbf{N}$ with $nC \leq d(g, h) \leq (n+1)C$. The proof is by induction on n. Suppose that $n = 0$. For $k, l \in \mathbf{Z}$, set $g_k = n_g - k[C/2]$ and $h_l = n_h - l[C/2]$. We now claim that *for all* $g_k \geq 0$ *there is an* h_l *such that*

$$|g_k - h_l| \leq [C/2] \text{ and } d(s(g)(g_k), s(h)(h_l)) \leq C.$$

To prove the claim, suppose first that $g_k \geq \max\{n_g, n_h\}$. Then there is a unique h_l with $0 \leq h_l - g_k \leq [C/2]$, and this h_l has the desired property. Next, suppose the claim is true for some $g_{k-1} \geq [C/2]$, so that $d(s(g)(g_{k-1}), s(h)(h_l)) \leq C$ for some h_l. There are two possibilities. If $g_k \leq h_l \leq g_{k-1}$ then since s is contracting we get $d(s(g)(g_k), s(h)(h_l)) \leq C$, provided $g_{k-1} \leq n_g$. But if $g_{k-1} > n_g$ then also $g_k \geq n_g$ so that $s(g)(g_{k-1}) = s(g)(g_k) = g$ and the claim is obviously true in this case too. If, on the other hand $g_{k-1} \leq h_l$, then under the further assumption that $h_l \leq n_h$, (2) gives $d(s(g)(g_{k-1}), s(h)(h_{l+1})) \leq C$ and we are back to the previous case. The same inequality is trivially true if $n_h < h_l$ since in this case $n_h \leq h_{l+1}$ and hence $s(h)(h_{l+1}) = s(h)(h_l) = h$. This proves the claim.

Next, let t, t' be natural numbers with $0 \leq t - t' \leq [C/2] \leq t$, and choose k so that $t - [C/2] \leq g_k \leq t$. Then the triangle inequality together with the claim give

$$d(s(g)(t - [C/2]), s(h)(t')) \leq d(s(g)(t - [C/2]), s(g)(g_k))$$
$$+ d(s(g)(g_k), s(h)(h_l)) + d(s(h)(h_l), s(h)(t'))$$
$$\leq [C/2] + C + 3[C/2] \leq 3C = \varphi(d(g, h))$$

This concludes the proof of the case $n = 0$.

Suppose inductively that whenever $d(u,v) \leq nC$, we have $d(s(u)(t - [C/2]), s(v)(t')) \leq 3nC$. Assume that $nC \leq d(g,h) \leq (n+1)C$ for some $n > 1$, and choose a point v at distance C from g along a geodesic joining g and h. Then $(n-1)C \leq d(v,h) \leq nC$ and

$$d(s(g)(t - [C/2]), s(h)(t')) \leq d(s(g)(t - [C/2]), s(v)(t'))$$
$$+ d(s(v)(t'), s(h)(t')) \leq (3 + 3n)C$$

This completes the proof.

2. The complex $P_d(G)$

For $d \in \mathbf{N}_+$, recall from [Gr] that there is a simplicial complex $P_d(G)$ whose set of vertices is G and whose simplices are all the finite subsets of G whose diameter is $\leq d$. $P_d(G)$ comes equipped with a simplicial action of G that is free and transitive on its set of vertices. This implies that the G-action on $P_d(G)$ has finite stabilizers. Note that if $d \leq d'$ there is an inclusion $P_d(G) \hookrightarrow P_{d'}(G)$. Moreover, $P_d(G)$ is G-equivariant in $P_{d'}(G)$.

The fact that balls in G (of any given radius and, say, centered at the identity element) are finite, implies that $P_d(G)$ is locally finite and finite dimensional, and that $P_d(G)/G$ is compact. The Theorem of Rips' quoted above is the assertion that, moreover, $P_d(G)$ is contractible for sufficiently large d.

The theorem of Ken Brown that we need can be formulated as follows (see [B]).

THEOREM [BROWN]. *Let X be a contractible G-complex such that the stabilizer of every cell is finitely presented and of type FP_∞. Let $\{X_j\}_{j \geq 1}$ be a filtration of X by G-equivariant subcomplexes, such that each X_j is finite mod G. Suppose that for every j there exists $j' \geq j$ such that $\pi_1(X_j) \to \pi_1(X_{j'})$, and $\tilde{H}_i(X_j) \to \tilde{H}_i(X_{j'})$ are trivial for all $i \geq 0$. Then G is finitely presented and of type FP_∞.*

We can now state

THEOREM 1. *Suppose that G admits a bounded combing, and let $d \in$ $\mathbf{N}$. Then there exists $d' \geq d$, such that the inclusion $P_d(G) \hookrightarrow P_{d'}(G)$*

induces trivial homomorphisms in homotopy and in reduced homology in all dimensions.

PROOF: It suffices to show that there is an integer $d' \geq d$ such that for every finite complex $K \leq P_d(G)$, the inclusion $K \hookrightarrow P_{d'}(G)$ is homotopic to the constant map collapsing K to the vertex $1 \in P_d(G)$. Such a homotopy will be obtained through a finite sequence of homotopies successively pushing K along the combing towards the vertex 1.

Suppose that G has a combing s as in (1'), and set $d' = \max\{\varphi(d), C\}$. Let K^0, the 0-skeleton of K, consist of $\{v_1, \ldots, v_m\}$ and, to simplify the notation, let n_i denote n_{v_i}, for $1 \leq i \leq m$. Suppose that, say, $n_1 = \max\{n_1, \ldots, n_m\}$. If $n_1 \leq [C/2]$ then $d(v_i, 1) \leq [C/2]$, so that $d(v_i, v_j) \leq C \leq d'$. It follows that K is contained in a simplex of $P_{d'}(G)$ whence it can be contracted to a point in this complex.

Suppose now that $n_1 > [C/2]$. For $p \geq 0$, we inductively define sets

$$K_p^0 = \{v_1^{(p)}, \ldots, v_m^{(p)}\}$$

(K_p^0 will be the 0-skeleton of a certain complex K_p inductively defined below) and functions $f_p \colon K_p^0 \to P_{d'}(G)$, so that $K_{p+1}^0 = f_p(K_p^0)$. To start the induction we set $v_i^{(0)} = v_i$, so that $K_0^0 = K^0$. Under the assumption that $n_1 = \max\{n_1, \ldots, n_m\}$ we define f_0 by $v_1^{(0)} \mapsto v_1^{(1)} = s(v_1)(n_1 - [C/2])$, and $v_j^{(0)} \mapsto v_j^{(1)} = v_j^{(0)}$ for all $j \geq 2$.

Suppose now that $K_p^0 = \{v_1^{(p)}, \ldots, v_m^{(p)}\}$ is already defined, where $v_i^{(p)} = s(v_i)(n_i - q_i[C/2])$ for some integer $q_i \geq 0$. Assume that, say, $n_1 - q_1[C/2] = \max\{n_1 - q_1[C/2], \ldots, n_m - q_m[C/2]\} > [C/2]$ (the process stops when this maximum is $\leq [C/2]$). We then define f_p by $v_1^{(p)} \mapsto v_1^{(p+1)} = s(v_1)(n_1 - (q_1 + 1)[C/2])$, and $v_j^{(p)} \mapsto v_j^{(p+1)} = v_j^{(p)}$ for all $j \geq 2$.

We will now prove inductively that (a) f_p can be extended to a simplicial map $f_p \colon K_p \to P_{d'}(G)$, where K_p is defined inductively to be $f_{p-1}(K_{p-1})$ when $p \geq 1$, and $K_0 = K$; and (b) f_p is homotopic to the inclusion $i_p \colon K_p \hookrightarrow P_{d'}(G)$. Both (a) and (b) will be a consequence of the following claim:

$$\text{if } d(v_i, v_j) \leq d, \text{ then } d(v_i^{(p+1)}, v_j^{(p+1)}) \leq d'.$$

The proof is by induction on p. Suppose that at stage p, $n_1 - q_1[C/2] = \max\{n_1 - q_1[C/2], \ldots, n_m - q_m[C/2]\} > [C/2]$ and that for some $j \neq 1$, $q_j > 0$. Then $n_1 - (q_1 + 1)[C/2] \leq n_j - q_j[C/2]$. Indeed, that $q_j > 0$ implies that v_j was moved along the combing towards the identity at some previous stage

$p' < p$. By the definition of f_p, this can only happen if $n_j - (q_j - 1)[C/2]$ was the largest parameter at stage p'. In particular, it must have been larger than the current largest parameter, i.e. $n_j - (q_j - 1)[C/2] \geq n_1 - q_1[C/2]$ as desired. Thus when $q_j > 0$, $(1')$ applies to give

$$d(v_1^{(p+1)}, v_j^{(p+1)}) = d(s(v_1)(n_1 - (q_1 + 1)[C/2]), s(v_j)(n_j - q_j[C/2]))$$
$$\leq \varphi(d(v_1, v_j)) \leq \varphi(d) \leq d'$$

When $q_j = 0$, put $t' = n_1 - q_1[C/2]$. Then $t' \geq n_j$ and

$$d(v_1^{(p+1)}, v_j^{(p+1)}) = d(s(v_1)(n_1 - (q_1 + 1)[C/2]), s(v_j)(t')) \leq d'$$

by $(1')$. To complete the proof of the claim, note that when $i \neq 1 \neq j$ $v_i^{(p+1)} = v_i^{(p)}$ and $v_j^{(p+1)} = v_j^{(p)}$, and the claim follows by induction.

To prove (a) we have to show that if $\sigma_p = \{v_{i_1}^{(p)}, \ldots, v_{i_r}^{(p)}\}$ is a simplex in K_p then $f_p(\sigma_p)$ is a simplex in $P_{d'}(G)$. This follows immediately from the claim, since σ_p is a simplex in K_p if and only if $\sigma = \{v_{i_1}, \ldots, v_{i_r}\}$ is a simplex in K.

For (b) we actually prove that f_p and the inclusion are simplicially homotopic, where two simplicial maps $f, g \colon M \to L$ are called *simplicially homotopic* if $f(\sigma) \cup g(\sigma)$ is contained in a simplex of L for every simplex σ of M. But the claim implies immediately that $\sigma_p \cup f_p(\sigma_p)$ is contained in a simplex of $P_{d'}(G)$ for all σ_p in K_p.

Since $C \geq 2$, a finite sequence of these homotopies will eventually put every vertex of K within $[C/2]$ of 1. Hence, if we write f_p as the composite $K_p \xrightarrow{\bar{f}_p} K_{p+1} \xrightarrow{i_{p+1}} P_{d'}(G)$, then we have a finite sequence

$$K_0 \xrightarrow{\bar{f}_0} K_1 \xrightarrow{\bar{f}_1} K_2 \xrightarrow{\bar{f}_2} \cdots \xrightarrow{\bar{f}_{N-1}} K_N$$

where all the vertices of K_N lie within $[C/2]$ of the identity. By (b), i_p is homotopic to $f_p = i_{p+1}\bar{f}_p$ for $0 \leq p \leq N - 1$. Hence i_0 is homotopic to $i_N \bar{f}_{N-1} \cdots \bar{f}_1 \bar{f}_0$, and by the first part i_N is homotopic to a constant map. This completes the proof.

THEOREM 2. *If a group admits a bounded combing then it is finitely presented and of type FP_∞.*

PROOF: Set $X = P_\infty(G)$, the "simplex" spanned by G, and let $X_j = P_j(G)$ for $j \in \mathbb{N}_+$. In view of Theorem 1, we may complete the proof by appealing to the Theorem of Brown quoted above.

COROLLARY 1. *Automatic groups are finitely presented and of type* FP_∞.

The next result includes Rips' Theorem.

THEOREM 3. *If a group G admits a contracting combing then there is a simplicial complex $P_d(G)$ with the following properties:*

(1) $P_d(G)$ *is contractible, locally finite and finite dimensional.*

(2) *The simplicial G-action on $P_d(G)$ is free and transitive on the set of vertices.*

(3) $P_d(G)/G$ *is compact.*

PROOF: Given a contracting combing s with constant C, set $d = C = d'$. Proceed now as in the proof of Theorem 1, replacing the claim in that proof by: if $d(v_i^{(p)}, v_j^{(p)}) \leq d$ then $d(v_i^{(p+1)}, v_j^{(p+1)}) \leq d$, which holds because s is contracting. The rest of the proof of Theorem 1 applies verbatim.

EXAMPLE: In particular, a torsion-free group satisfying the hypothesis of Theorem 3 must have finite cohomological dimension. Thus the finitely presented, infinite dimensional group of type FP_∞ studied in [BG], being torsion-free, cannot admit a contracting combing. It is not known whether it admits a bounded one.

3. Hyperbolic combings

In this section we characterize the type of combings possessed by hyperbolic groups. Recall ([BGHHSST], [CDP], [S]) that G is δ-hyperbolic if

$$(4) \quad d(z,u) + d(v,w) \leq \max \{d(z,v) + d(u,w), d(z,w) + d(u,v)\} + 2\delta$$

for all $z, u, v, w \in G$.

We also need a characterization, due to Gromov, of hyperbolic groups in terms of *isoperimetric inequalities*. Let G be a finitely presented group and let K be a finite polyhedron with fundamental group isomorphic to G. Let γ be a circuit in the 1-skeleton of the universal cover $\tilde{K}$ of K. The length of γ, $\ell(\gamma)$, is the number of 1-simplices in γ. Since $\tilde{K}$ is 1-connected, there is a simplicial disc with γ as boundary. Let $\alpha(\gamma)$ denote the minimal area, i.e. the minimal number of 2-simplices in such a disc. We then say that G satisfies a *linear* [resp. *quadratic*] isoperimetric inequality if there is a constant C such that for all γ as above, one has $\alpha(\gamma) \leq C\ell(\gamma)$ [resp. $\alpha(\gamma) \leq C\ell(\gamma)^2$]. This property is independent of the choice of K. See [Ge1] [AB] for more details.

THEOREM [GROMOV]. *If G is a finitely presented group, then the following conditions are equivalent:*

(i) *G is a hyperbolic group.*
(ii) *G satisfies a linear isoperimetric inequality.*

REMARK: See [CDP] for the proof of $(i) \implies (ii)$. H. Short has given a very nice proof of the more delicate converse. See [S] or [GS].

We can now prove

THEOREM 4. *(i) Let G be a hyperbolic group. Then for all $r \in (1/2, 1)$ there is a $C \geq 2$ such that G admits a C-r-contracting combing.*

(ii) Suppose that G is a group admitting a C-r-contracting combing, for some $C \geq 7$ and $1/2 < r \leq 3/5$. Then G satisfies a linear isoperimetric inequality.

REMARK: It follows that G is hyperbolic if and only if it admits a C-r-contracting combing as in *(ii)*.

PROOF: (i) Suppose that G is δ-hyperbolic with respect to a set of generators A. Choose a geodesic combing s, and given $r \in (1/2, 1)$, take any C satisfying $C \geq (1 + 2\delta)/(r - 1/2)$. Given $g, h \in G$, $t \in [[C/2], n_g]$, $k \in [2, C]$ and t' with $t' \leq t$, suppose that $d(s(g)(t), s(h)(t')) \leq k$. Apply (4) to the points $s(h)(t')$, $s(g)(t - [k/2])$, $s(g)(t)$, and 1, and subtract $d(s(g)(t), 1) = t$ from both sides to get

$$d(s(g)(t - [k/2]), s(h)(t')) \leq \max \{k - [k/2], [k/2] + t' - t\} + 2\delta$$
$$\leq [k/2] + 1 + 2\delta \leq [k/2] + (r - 1/2)C$$

as desired.

(ii) Suppose that G admits a C-r-contracting combing s as above. Then $P_C(G)$ is contractible by Theorem 3, and since G acts on it with finite stabilizers, this complex can be used to check that G satisfies a linear isoperimetric inequality (see [AB]). As in [Gr] (see also [CDP] or [S]), we prove the following more precise result. Let S^1 be a simplicial circle with N_1 edges, and let $f : S^1 \to P_C(G)$ be a simplicial map. Then there is a simplicial disk D^2 with N_2 triangles satisfying $N_2 \leq 3N_1$, $\partial D^2 = S^1$, and such that f extends to a simplicial map $D^2 \to P_C(G)$.

The proof is by induction on N_1. Let $\{v_1, \ldots, v_q\}$ be the set of vertices of the curve $f(S^1)$, and suppose that v_1 is a vertex with maximal n_j (as usual we abbreviate $n_{v_j} = n_j$). Let v_2, v_3 be the vertices on the curve, adjacent to v_1 (thus $d(v_1, v_i) \leq C$, $i = 2, 3$). Without loss of generality, we assume that $n_1 \geq n_2 \geq n_3$, and consider several cases.

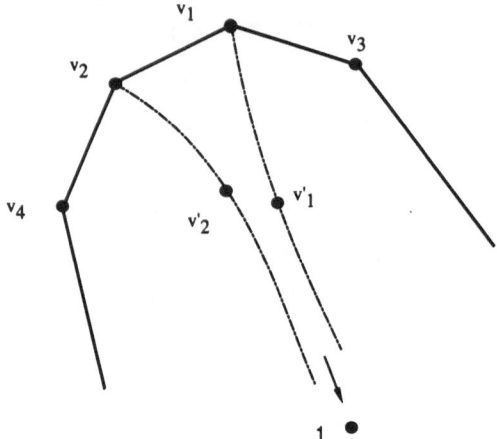

Case I: $n_2 \leq [C/2]$. Then $d(v_2, v_3) \leq C$. Replacing the edges $\overline{v_2 v_1} \cup \overline{v_1 v_3}$ by $\overline{v_2 v_3}$ gives a curve of length $N_1 - 1$. To complete the proof of this case, add the triangle (in $P_C(G)$!) $\triangle(v_2 v_1 v_3)$, to the simplicial disk provided by the induction hypothesis.

From now on we assume that $n_2 > [C/2]$.

Case II: $n_1 = n_2$. We consider two subcases.

(a) $n_1 = n_2 = n_3$. Set $v_2' = s(v_2)(n_2 - [C/2])$. Then by (3) $d(v_2', v_1) \leq rC$ and $d(v_2', v_4) \leq rC$. Now set $v_1' = s(v_1)(n_1 - [[rC]/2])$. Applying (3) again, $d(v_2', v_1') \leq (r - 1/2)C + [[rC]/2] \leq (3r - 1)C/2$. Similarly, if we let $v_3' = s(v_3)(n_3 - [[rC]/2])$, then $d(v_3', v_1') \leq (3r - 1)C/2$ and $d(v_3', v_5) \leq rC$. Hence

$$d(v_1', v_4) \leq d(v_1', v_2') + d(v_2', v_4) \leq (5r - 1)C/2 \leq C.$$

One shows similarly that $d(v_1', v_2)$, $d(v_1', v_5)$ and $d(v_1', v_3)$ are all $\leq C$. We then replace the edges $\overline{v_4 v_2} \cup \overline{v_2 v_1} \cup \overline{v_1 v_3} \cup \overline{v_3 v_5}$ by $\overline{v_4 v_1'} \cup \overline{v_1' v_5}$ to get a curve with $N_1 - 2$ edges. We then add the triangles $\triangle(v_4 v_1' v_2)$, $\triangle(v_2 v_1' v_1)$, $\triangle(v_1 v_1' v_3)$ and $\triangle(v_3 v_1' v_5)$ to the simplicial disk (with $\leq 3(N_1 - 2)$ triangles) given by induction. The new disk thus obtained has at most $3(N_1 - 2) + 4 < 3N_1$ triangles as desired.

(b) $n_1 = n_2 > n_3$. Let v_1' and v_2' be as in (a). If we set $v_1'' = s(v_1)(n_1 - [C/2])$ then $d(v_1'', v_3) \leq rC$. On the other hand, $2d(v_1', v_1'') = 2([C/2] - [[rC]/2]) \leq C - [rC] + 1 \leq 2 + C/2$. Thus $d(v_1', v_3) \leq d(v_1', v_1'') + d(v_1'', v_3) \leq 1 + 17C/20 \leq C$ for $C \geq 7$. We then replace $\overline{v_4 v_2} \cup \overline{v_2 v_1} \cup \overline{v_1 v_3}$ by $\overline{v_4 v_1'} \cup \overline{v_1' v_3}$, and add the triangles $\triangle(v_4 v_1' v_2)$, $\triangle(v_2 v_1' v_1)$ and $\triangle(v_1 v_1' v_3)$.

Case III. $n_1 > n_2$. As before, we consider two subcases:

(a) $n_2 \leq n_1 - [C/2]$. Set $v_1' = s(v_1)(n_1 - [C/2])$ and $v_1'' = s(v_1)(n_1 - [C/2] - [[rC]/2])$. Then (3) gives $d(v_i, v_1') \leq rC$ and $d(v_i, v_1'') \leq (3r - 1)C/2$ for $i = 2, 3$. Thus $d(v_2, v_3) \leq (3r - 1)C < C$, and we are back in the situation considered in case I.

(b) $n_1 - [C/2] < n_2$. Let $v_1' = s(v_1)(n_1 - [C/2])$, $v_2' = s(v_2)(n_2 - [[rC]/2])$ and $v_2'' = s(v_2)(n_2 - [C/2])$. Proceeding as in case II(b) above, use $C \geq 7$ to get $d(v_2', v_4) \leq C$, and notice that also $d(v_2', v_1) \leq C$. Finally, replace $\overline{v_4 v_2} \cup \overline{v_2 v_1} \cup \overline{v_1 v_3}$ by $\overline{v_4 v_2'} \cup \overline{v_2' v_3}$, and then add the triangles $\triangle(v_4 v_2' v_2)$, $\triangle(v_2 v_2' v_1)$ and $\triangle(v_1 v_2' v_3)$. This completes the proof of the subcase and that of the Theorem.

4. Quadratic isoperimetric inequality

For completeness, we include a proof that a group with a combing of any of the types considered in this paper will satisfy a quadratic isoperimetric inequality, *provided* that the combing is *quasi-geodesic*. Call a combing s quasi-geodesic if there are constants $C > 0$ and $D \geq 0$ such that for all $g \in G$ and $t_1, t_2 \in [0, n_g]$,

$$|t_1 - t_2| \leq Cd(s(g)(t_1), s(g)(t_2)) + D$$

Geodesic combings are quasi-geodesic. By [CEHPT, Thm. 9.5] automatic groups admit quasi-geodesic combings. Thus Theorem 5 below applies to automatic groups. Of course, we already know (Theorem 4) that groups admitting a C-r-combing (for appropriate C and r) satisfy a linear isoperimetric inequality.

Before giving its proof, which is due to Thurston (see also [BGSS]), we need a reformulation of the geometric definition of isoperimetric inequality given in §3. Starting with a finite presentation $\mathcal{P} = \langle X | R \rangle$ of a group G, define a finite complex $K(\mathcal{P})$ as follows. Form a bouquet of 1-spheres, one for each element of X, and for each relator in R attach a 2-cell by a map spelling out the given relator. Using this complex and the definition in §3, one can obtain the following algebraic interpretation of isoperimetric inequalities. If γ is a freely reduced word of length $\ell(\gamma)$ in $F(X)$, the free group on X, and γ represents 1 in G, note there are words g_i, relators r_i and $\epsilon_i = \pm 1$ such that

$$\gamma = \prod_{i=1}^{m} g_i r_i^{\epsilon_i} g_i^{-1} \text{ in } F(X).$$

Then G satisfies a linear [resp. quadratic] isoperimetric inequality if there is a constant C such that every such γ can be written as above with an m satisfying $m \le C\ell(\gamma)$ [resp. $m \le C\ell(\gamma)^2$].

THEOREM 5. *Suppose that a group G admits a quasi-geodesic combing of any of the three types considered in §1. Then G satisfies a quadratic isoperimetric inequality.*

REMARKS: (a) It can be shown that this implies that G has a solvable word problem (see [Ge1] [S]).

(b) The group $\mathbf{Z} \times \mathbf{Z}$ is not hyperbolic [Gr], but it is automatic [CEHPT]. Thus it satisfies a quadratic but not a linear, isoperimetric inequality.

(c) By the Theorem, groups that admit a contracting combing satisfy (at least) a quadratic isoperimetric inequality, but I don't know if they must actually satisfy a linear one.

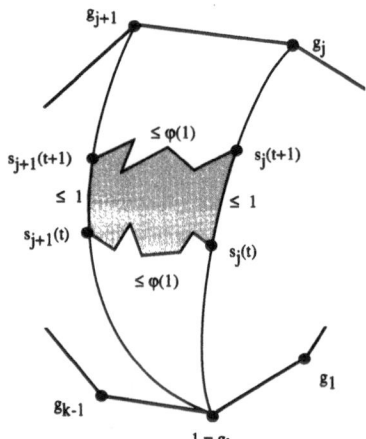

PROOF: Let s be a quasi-geodesic combing of G with respect to a finite set of generators X. By §1 it suffices to assume that s is bounded. Let $G = F(X)/N$ and suppose that $\gamma \in N$ is a freely reduced word of length k, say $\gamma = x_{i_1} \ldots x_{i_k}$, $x_{i_j} \in X$. This gives a circuit in the Cayley graph of G with respect to X, with the following elements of G as vertices: $g_1 = x_{i_1}$, $g_2 = x_{i_1} x_{i_2}, \ldots, g_k = x_{i_1} x_{i_2} \ldots x_{i_k} = 1$. Note that for $0 \leq j \leq k-1$, $g_{j+1} = g_j x_{i_{j+1}}$ (where $g_0 = 1$). Set $s_j = s(g_j)$ and $n_j = n_{g_j}$. Since s is quasi-geodesic, we have

$$n_j \leq Cd(1, s_j(n_j)) + D \leq \frac{C\ell(\gamma)}{2} + D$$

for all $1 \leq j \leq k$. Thus we can show that $g_{j+1} = g_j x_{i_{j+1}}$ using no more than $\max\{n_j, n_{j+1}\} \leq Ck/2 + D$ relators, each of length $\leq 2\varphi(1) + 2$ (here φ is as in (1)). Hence showing that $\gamma = 1$ requires no more than

$$k(\frac{Ck}{2} + D) \leq Ak^2, \qquad \text{for some } A > 0$$

relators (each of length $\leq 2\varphi(1) + 2$) as desired. This completes the proof.

REMARK: Note that it follows from the above proof that

$$R = \{\gamma \in F(X) | \gamma = 1 \text{ in } G, \text{ and } \ell(\gamma) \leq 2\varphi(1) + 2\}$$

is a finite set of relators of G. This gives an alternative proof, valid in case the bounded combing is quasi-geodesic, that G is finitely presented.

References

[AB] J.M. Alonso and S.G. Brick, *On isoperimetric inequalities and products of groups*, (In preparation).

[B] K.S. Brown, *Finiteness properties of groups*, J. Pure Appl. Algebra **44** (1987), 45–75.

[BG] K.S. Brown and R. Geoghegan, *An infinite-dimensional torsion-free FP_∞ group*, Invent.Math. **77** (1984), 367–381.

[BGHHSST] W. Ballmann, E. Ghys (ed), P. de la Harpe (ed), E. Salem, R. Strebel et M. Troyanov, *Sur les groupes hyperboliques d'après Mikhael Gromov*, (preprint) (1989).

[BGSS] G. Baumslag, S. M. Gersten, M. Shapiro and H. Short, *Automatic groups and amalgams*, (preprint) (1989).

[CEHPT] J.W. Cannon, D.B.A. Epstein, D.F. Holt, M.S. Paterson, and W.P. Thurston, *Word processing and group theory*, (preprint) (1988).

[CDP] M.Coornaert, T. Delzant, et A. Papadopoulos, *Notes sur les groupes hyperboliques de Gromov*, preprint (1989).

[Ge1] S. Gersten, *Dehn functions and l_1-norms of finite presentations*, preprint (1989).

[Ge2] S. Gersten, (private communication) (December 1989).

[Gr] M. Gromov, *Hyperbolic groups*, "Essays in group theory," S.M. Gersten (ed.), M.S.R.I Publications; 8, 1987, pp. 75–263.

[GS] S. Gersten and H. Short, *Small cancellation theory and automatic groups*, M.S.R.I. preprint # 03323-89 (1989).

[S] H. Short (ed.), *Notes on negatively curved groups*, M.S.R.I. preprint # 08023-89 (1989).

Department of Mathematics, University of Stockholm, P.O. Box 6701, S-113 85 Stockholm, SWEDEN.
Current address: Department of Mathematics, Cornell University, White Hall, Ithaca, N.Y. 18453, USA.

Automatic Groups and Amalgams — A Survey

G. Baumslag, S.M. Gersten, M. Shapiro, and H. Short

1. Basic objective

Automatic and aysnchronously automatic groups were invented by J.W. Cannon, D.B.A. Epstein, D.F. Holt, M.S. Patterson and W.P. Thurston in [CEHPT] a few years ago. The primary objective of this paper is to report, without proofs, on a number of new results about such groups. (Proofs will appear elsewhere, in due course.) Most of these results are concerned with the construction of new automatic and asynchronously automatic groups from old by means of amalgamated products. Our secondary objective here is to survey a little of the theory of automatic groups and the like, keeping the account reasonably self-contained.

2. Finite state automata

Let $\mathcal{A}$ be the finite set $\{a_1, \ldots, a_q\}$ and let $\mathcal{A}^*$ be the set of all *words*

$$w = a_1 \ldots a_n$$

made up from *letters* in $\mathcal{A}$ (including the empty word e); we term n *the length of w*, which we denote by $\ell(w)$. This set $\mathcal{A}^*$ together with the binary operation concatenation (i.e. juxtaposition) is a monoid, which we will make use of throughout. The subsets of $\mathcal{A}^*$ are often referred to as *languages over $\mathcal{A}$* . We shall be concerned with special languages over $\mathcal{A}$ termed *regular languages over $\mathcal{A}$* or sometimes *regular sets (over $\mathcal{A}$)*. These depend for their definition, which will be given shortly, on the notion of a finite state automaton.

A *finite state automaton* is a quintuple $M = (S, Y, \mathcal{A}, \tau, s_0)$, where

(i) S is a finite set of *states*;

(ii) Y is a subset of S, the set of *accept states*;

(iii) $\mathcal{A}$ is a finite set, the *alphabet*, the elements of which are *letters*;

(iv) τ is a function from $S \times \mathcal{A} \to S$, the *transition function*;

(v) s_0 is an element of S, the *initial state* or *start state*.

M can be thought of as a machine with a head that scans a tape, which is in a horizontal position stretching from left to right. The tape is divided up into a finite number of squares. Each square has a letter printed on it. The left-most part of the tape is fed into M, which starts up in the initial state s_0. M reads the first letter on the tape, and then moves on to the second letter and the machine goes into a new state. This new state is determined by the transition function τ and the letter that is being scanned by M. The process continues until the last letter on the tape is read. The machine then goes into a new state and stops. If the last state is an accept state, then the string of letters on the tape is *accepted* by M. Otherwise it is rejected. We define *the language recognized by M* or *the language accepted by M* or *the language of M* to be the set $L(M)$ of words accepted by M. The language $L(M)$ of M can therefore be described as follows. Given a word $w = a_1 \ldots a_n$, let $t_0 = s_0$ and let $t_i = \tau(t_{i-1}, a_i)$ $(i = 1, \ldots, n)$. Then

$$L(M) = \{w = a_1 \ldots a_n \mid t_n \in Y\}.$$

A language recognised by a finite state automaton is termed a *regular language* or a *regular set*.

The reader who is not familiar with these notions might find it instructive to think about the following example of a finite state automaton M. The set S of states of M consists of s_0, the initial state, s_1, s_2, f, the set Y of accept states of M consists of s_0, s_1, s_2, the alphabet $\mathcal{A}$ of M consists of x, X and the transition function τ of M is defined as follows:

$$\tau(s_0, x) = s_1, \tau(s_0, X) = s_2, \tau(s_1, x) = s_1,$$
$$\tau(s_1, X) = f, \tau(s_2, x) = f, \tau(s_2, X) = s_2.$$

Then it is not hard to see that the language $L(M)$ of M consists of all those words in x and X which do not contain a pair of consecutive letters of either the form xX or Xx.

For more information on automata and languages, we refer the reader to the book by Hopcroft and Ullman [HU].

3. Padded alphabets

We shall be concerned with pairs of elements of $\mathcal{A}^\star$, i.e. with the direct product $\mathcal{A}^\star \times \mathcal{A}^\star$ of two copies of the monoid $\mathcal{A}^\star$. A little thought reveals that because $\mathcal{A}^* \times \mathcal{A}^*$ is not generated by $\mathcal{A} \times \mathcal{A}$, there is a slight problem in finding an appropriate alphabet for the "product" of two languages over $\mathcal{A}$. In order to get around this problem, we introduce a so-called *padding symbol* \$, which by assumption is not contained in $\mathcal{A}$. We define

$$(\mathcal{A} \cup \{\$\}) \times (\mathcal{A} \cup \{\$\}) - \{(\$,\$)\}$$

to be the *padded alphabet* of $\mathcal{A}$ and think of it as a subset of the direct product of two copies of $(\mathcal{A} \cup \{\$\})^*$. Notice that the padded alphabet of $\mathcal{A}$ freely generates a free submonoid of this direct product. It is in this monoid that our calculations take place. We define a mapping ν of $\mathcal{A}^* \times \mathcal{A}^*$ into $(\mathcal{A} \cup \{\$\})^* \times (\mathcal{A} \cup \{\$\})^*$ as follows. If $u = a_1 \ldots a_n, v = b_1 \ldots b_m$, then

$$\nu : (u,v) \longrightarrow (a_1,b_1) \ldots (a_m,b_m)(a_{m+1},\$) \ldots (a_n,\$) \; if \; m < n,$$

$$\nu : (u,v) \longrightarrow (a_1,b_1) \ldots (a_n,b_n) \; if \; m = n,$$

$$\nu : (u,v) \longrightarrow (a_1,b_1) \ldots (a_n,b_n)(\$,b_{n+1}) \ldots (\$,b_m) \; if \; m > n$$

and

$$\nu : (e,e) \longrightarrow e.$$

4. Monoid generating sets

Now let G be a group. Then we term $\mathcal{A}$ *a set of monoid generators of G* if it comes equipped with a map, termed *a monoid generating map*,

$$a \mapsto \bar{a} \; (a \in \mathcal{A}, \; \bar{a} \in G)$$

of $\mathcal{A}$ into G whose extension μ to a monoid homomorphism of $\mathcal{A}^*$ into G is a surjection. This means that every element of G can be expressed as a product of the elements in $\{\overline{a_1}, \ldots, \overline{a_q}\}$. The map $a \mapsto \bar{a}$ need not be, but often is, monic, indeed even an inclusion. However it is important to note, in order to avoid confusion, that in many cases $\mathcal{A}$ is not a subset of G. The image of $w \in \mathcal{A}^*$ under μ is usually denoted here by $\bar{w}$.

5. Automatic groups

Suppose that G is a group and that $\mathcal{A}$ is a finite set of monoid generators of G. A regular subset L of $\mathcal{A}^{\star}$ which maps surjectively to G via the map μ is termed a *regular language (over $\mathcal{A}$) for G* and, following Gilman [Gi], we term the pair $(\mathcal{A}, L)$ a *rational structure for G*. The word $w \in L$ is referred to as a *representative of $g \in G$* if $\overline{w} = g$.

Suppose that $(\mathcal{A}, L)$ is a rational structure for the group G. We need to consider the following subsets of $\nu(\mathcal{A}^{\star} \times \mathcal{A}^{\star})$:

$$L_{=} = \{\nu((w_1, w_2)) \mid w_1, w_2 \in L, \overline{w_1} = \overline{w_2}\},$$

and, for each $w \in \mathcal{A}^{\star}$ the set

$$L_w = \{\nu((w_1, w_2)) \mid w_1, w_2 \in L, \overline{w_1} = \overline{w_2 w}\}.$$

In the event that $w = a_i \in \mathcal{A}$ we sometimes denote L_w by L_i and term it the *multiplication checker for a_i* or the $i - th$ comparator. We will refer to $L_{=}$ as the *equality checker*.

The group G is termed an *automatic group* if there exists a rational structure $(\mathcal{A}, L)$ for G with the following properties:

(i) the equality checker $L_{=}$ is regular;

and

(ii) the multiplication checker for a_i, is regular for each $i = 1, 2, \ldots, q$.

We term such a regular language L for G *automatic* or *automatic over $\mathcal{A}$*. Thus a group G is automatic if it has a rational structure $(\mathcal{A}, L)$ for G where L is automatic. In other words a group G is an automatic group if there is a regular language L for G such that

(i) there is a finite state automaton, termed *the equality checker*, which "checks" whether two words in L represent equal elements in G;

and

(ii) there is, for each i, a finite state automaton, termed *the comparator automaton for a_i* which checks whether any pair of words in L differ by right multiplication by a_i.

The infinite cyclic group G is an automatic group. In order to see that this is so, suppose that x generates G and put $X = x^{-1}$. Let $\mathcal{A} = \{x, X\}$. Then

$\mathcal{A}$ is a set of monoid generators of G (with generating map "inclusion"). We have to find a rational structure $(\mathcal{A}, L)$ for G with L automatic. We have already chosen $\mathcal{A}$. We take $L = L(M)$, where M is the solitary example of a finite state automaton that we gave in §2. Thus L is nothing but the set of all reduced words in $\{x\}$ and so the map μ is monic in this instance. It follows that $L_=$ is the diagonal of L, i.e. the set of all pairs $(w, w) \mid w \in L$. We need to prove that this set and the sets L_x and L_X are all regular. This requires a little thought – we leave it to the reader to construct the finite state automata which recognise these languages.

It is useful to note that if G is an automatic group and if $\mathcal{B}$ is *any* finite set of monoid generators for G, then there is an automatic language for G over $\mathcal{B}$ ([CEHPT]). Thus the property of being automatic is independent of the choice of the finite set of monoid generators.

6. Some examples and properties of automatic groups

As we have already noted, automatic groups were invented by J.W. Cannon, D.B.A. Epstein, D.F. Holt, M.S. Patterson and W.P. Thurston in [CEHPT], although the notion was hinted at in earlier work of J.W. Cannon [C]. Their very definition suggests that they have a reasonably simple structure. This is manifested by the fact that automatic groups are finitely presented and also that they have a solvable word problem. These results are due to Cannon et al. [CEHPT]. On the other hand automatic groups can be quite complicated. For instance they include the so-called *hyperbolic* groups of M. Gromov [Gr], which, following a suggestion of J.W. Cannon, we shall term here *negatively curved* groups and which we will define below. In particular the fundamental groups of closed hyperbolic manifolds are negatively curved (Gromov [Gr]) and hence are automatic. Moreover Cannon et al. [CEHPT] have proved that many common finitely presented groups are automatic; for example all *finite groups*, all *finitely generated free groups* and all *finitely generated abelian groups* are automatic. More generally they have shown that *the class of automatic groups is closed under finite free products, finite direct products and finite extensions* (but not under finitely generated subgroups. Somewhat surprisingly a finitely generated nilpotent group is automatic if and only if it contains a subgroup of finite index which is abelian [CEHPT]. Recently Gersten and Short [GS1], [GS2] have shown that all of the non-metric small cancellation groups of Lyndon (see [LS]) are also automatic.

It is clear then that automatic groups constitute a very interesting class of groups.

As we have already seen, an automatic group is described by an automatic language over an alphabet. This language is the langauge recognized by a finite state automaton. Epstein and Rees [ER] have written computer programs to compute this finite state automaton. Thus it is in some sense practical to use computers to actually compute "products" in automatic groups. This means that Epstein and Rees have built a bridge between part of combinatorial group theory and computer science, which may well have very interesting consequences.

7. Two tape automata

Cannon et al [CEHPT] have also introduced the notion of an asynchronous automatic group which is based on a *two tape automaton* . These two tape automata were introduced by Rabin and Scott [RS].

DEFINITION: Let $\mathcal{A}$ be a finite alphabet and let ϵ be a symbol not in $\mathcal{A}$. A *two tape automaton* T over the alphabet $\mathcal{A}$ is a finite state automaton $M = (\mathcal{B}, S, Y, \tau, s_0)$, where $\mathcal{B} = \mathcal{A} \cup \{\epsilon\}$, together with a partition $S = S_1 \sqcup S_2$ of S.

We can associate to each word $w = a_1 \ldots a_n \in L$, the language accepted by M, a pair $\Phi(w) = (u, v) \in \mathcal{B}^\star \times \mathcal{B}^\star$ as follows. Let $s_0 = t_0, t_1, \ldots, t_n$ be the succession of states that arise when M accepts w: i.e. $t_1 = \tau(t_0, a_1), t_2 = \tau(t_1, a_2), \ldots, t_n = \tau(t_{n-1}, a_n)$.

Then u is the word obtained from $w = a_1 \ldots a_n$ by omitting the letters a_j for which $t_{j-1} \in S_2$, and the word v is obtained from w by omitting the letters a_i for which $t_{i-1} \in S_1$. Notice that after time t, M has read the initial segment $w(t) = a_1 \ldots a_t$ and hence there are monotonic reparameterization functions ϕ_w, ψ_w such that $\Phi(w(t)) = (v(\phi_w(t)), v(\psi_w(t)))$ $(\phi_w(t) \leq t)$.

The *language* $\Lambda \subset \mathcal{A}^\star \times \mathcal{A}^\star$ of the two tape automaton T is the set of all pairs $(u', v') \in \mathcal{A}^\star \times \mathcal{A}^\star$ such that $(u'\epsilon, v'\epsilon) = \Phi(w)$ for some $w \in L$.

We shall sometimes think of T as a machine with two heads, scanning two separate tapes, say T_1 and T_2. Then T, which of course has the same set of states as M, *accepts* a pair of input words $(u'\epsilon, v'\epsilon)$, printed respectively on the tapes T_1 and T_2, in the following way (any other sequence of events leads the machine to reject a pair of input words). According to the state s it is in, the machine reads a letter from the tape T_1 (if $s \in S_1$) or the tape T_2 (if $s \in S_2$). After reading a letter (other than an ϵ) from the appropriate tape, the machine changes state, according to τ, and proceeds to scan the next position on the tape from which it has read, while continuing to scan the same position as before on the other tape. When the machine reads a symbol ϵ for the first time, from the tape T_i, the head does not advance on T_i, but the new state to which the machine goes lies in S_{3-i} (i.e. is not in S_i), and all subsequent letters are read from the other tape. The second time a symbol ϵ is read, necessarily from the other tape, all letters will have been read, and the machine will be in an accept state.

8. Asynchronously automatic groups

We have already noted that Cannon et al [CEHPT] have also introduced the notion of an *asynchronously automatic group*. The definition is analogous to that of an automatic group. More precisely the group G is termed *asynchronously automatic* if it has a rational structure $(\mathcal{A}, L)$ such that both the equality checker

$$L_{(=)} = \{(u, v) \mid u, v \in \mathcal{A}^*, \ \overline{u} = \overline{v}\}$$

and the multiplication checkers

$$L_{(i)} = \{(u, v) \mid u, v \in \mathcal{A}^*, \ \overline{u} = \overline{va_i}\}$$

are all languages of two tape automata. It turns out that if $L_=$ is the language of a finite state automaton, then $L_{(=)}$ is the language of a two tape automaton and similarly for the L_i. This implies that the class of automatic groups is contained in the class of aysnchronously automatic groups. It also turns out, as in the case of automatic groups, that asynchronously automatic groups are finitely presented and have solvable word problem ([CEHPT]. They play a very important role in our work.

9. Cayley graphs and fellow travellers

Let $\mathcal{A}$ be a finite set of monoid generators of the group G. The *Cayley graph of G with respect to $\mathcal{A}$* is the directed labelled graph $\Gamma_{\mathcal{A}}(G)$ with

(1) a vertex labelled g for each element $g \in G$;

(2) for each $a \in \mathcal{A}$, and for each $g \in G$, there is a directed edge labelled a from the vertex labelled g to the vertex labelled $g\bar{a}$ and a directed edge labelled a^{-1} from the vertex labelled $g\bar{a}$ to the vertex labelled g. Notice here that the labels a^{-1} that we have introduced here can be thought of as constituting a second set $\mathcal{A}^{-1}$ in one-to-one correspondence with $\mathcal{A}$.

The Cayley graph is always path connected.

If P and Q are vertices in a connected, directed graph Γ, then a path γ from P to Q is a finite sequence of edges $e_1, \ldots, e_n$ such that the origin $o(e_1)$ of e_1 is P, the terminus $t(e_n)$ of e_n is Q and $t(e_i) = o(e_{i+1})$ for $i = 1, \ldots, n-1$. The length of such a path γ is defined to be n. The distance $d(P, Q)$ between two vertices P and Q is defined to be the minimum length of a path from P to Q if $P \neq Q$ and 0 if $P = Q$. This turns Γ into a metric space. In particular then $\Gamma_{\mathcal{A}}(G)$ can be viewed as a metric space. A shortest path from the group element g to the group element h is termed $geodesic$ and a "triangle" in $\Gamma_{\mathcal{A}}(G)$ is termed a $geodesic$ $triangle$ if its sides are geodesics. We shall have occasion to refer to such geodesic triangles when we define negatively curved groups.

We now define $\overline{a^{-1}} = \bar{a}^{-1}$ $(a \in A)$ and define $\bar{w}$ for each word $w = b_1 \ldots b_n \in (\mathcal{A} \cup \mathcal{A})^\star$ by $\bar{w} = \overline{b_1} \ldots \overline{b_n}$. Then each word $w = b_1 \ldots b_n \in (\mathcal{A} \cup \mathcal{A})^\star$ can be turned into a map from the set of non-negative integers $0, 1, \ldots$ into G by setting $w(t) = \overline{b_1 \ldots b_t}$ $(t \leq n)$, $w(t) = \bar{w}$ $(t > n)$. We then term two words $u, v \in \mathcal{A}$ k-$fellow$ $travellers$ if $d(u(t), v(t)) \leq k$ for all t in $\Gamma_{\mathcal{A}}(G)$. We term u, v $asynchronous$ k-$fellow$ $travellers$ if we can find monotonic functions ϕ and ψ so that $d(u(\phi(t)), v(\psi(t))) \leq k$ in $\Gamma_{\mathcal{A}}(G)$.

10. Regular subgroups and amalgamated products

Let A be a group and let C be a subgroup of A. Suppose that there exists a rational structure $(\mathcal{A}, L_{\mathcal{A}})$ for A with the added property that there is a regular language J over $\mathcal{A}$ contained in $L_{\mathcal{A}}$ which maps via μ onto C. Then C is termed a $regular$ or $L_{\mathcal{A}}$-$regular$ $subgroup$ of A. Gilman [Gi] has proved that such regular subgroups are finitely generated. Suppose now that we denote the set of right cosets aC of C in A by A/C. We term a regular set M contained in $L_{\mathcal{A}}$ a $regular$ $language$ $with$ $uniqueness$ for A/C if the mapping

$$w \mapsto \bar{w}C \ (w \in M)$$

is a bijection between M and A/C.

All of our theorems about amalgamated products will be proved by appealing to the following very general theorem.

THEOREM 1. *Let G be the generalised free product of the automatic groups A and B amalgamating C:*

$$G = A \star_C B.$$

Let $\mathcal{A}$ be a finite set of monoid generators for A, let $\mathcal{B}$ be a finite set of monoid generators for B, let L_A be an automatic language for A over $\mathcal{A}$ and let L_B be an automatic language for B over $\mathcal{B}$. Suppose that the following conditions hold:

(1) *C is regular with respect to L_A (and hence there is a regular language $L(C) \subseteq L_A$ with exactly one representative for each element of C);*

(2) *there is a regular language $L(A/C)$ (contained in L_A) with uniqueness for A/C and a regular language $L(B/C)$ (contained in L_B) with uniqueness for B/C;*

(3) *there is a constant k such that whenever $u \in L(C)$ and $v \in L_B$ represent the same element of C, then u and v are k-fellow travellers (asynchronous k-fellow travellers) in $\Gamma_{A \cup B}(G)$;*

(4) *if $u \in L(A/C)$ (respectively $L(B/C)$), $v \in L(C)$ and $w \in L_A$ (respectively L_B) and if $\overline{uv} = \overline{w}$, then uv and w are k-fellow travellers (asynchronous k-fellow travellers).*

Then G is (asynchronously) automatic.

The assertion in (1) that there is a regular language $L(C)$ for C with exactly one representative for each element of C follows from general principles and we shall not attempt to justify it here.

There is an analogous theorem which holds also when A and B are asynchronously automatic.

11. Amalgamated products of abelian groups

We noted earlier that a finitely generated abelian group is automatic. Moreover it follows readily from the fact that a subgroup of a finitely generated abelian group is a direct factor of a subgroup of finite index that subgroups of finitely generated abelian groups are L_A-regular for an appropriate choice of $\mathcal{A}$ and L_A. This means that we have verified that at least some of the hypothesis of Theorem 1 is in effect when A and B are finitely generated abelian groups. Indeed it is not hard to deduce the following theorem from Theorem 1.

THEOREM 2. *Every amalgamated product of two finitely generated abelian groups is automatic.*

So it follows, in particular, that the groups

$$G_{m,n} = < a, b; a^m = b^n >$$

(and hence also all torus knot groups) are automatic.

12. Negatively curved groups, Dehn functions, and isoperimetric inequalities

Suppose next that $\langle X; R \rangle$ is a presentation of the group G. Thus the free group F on X comes equipped with a homomorphism onto G with kernel K the normal closure in F of R. Following our earlier convention with monoids we denote the image of $w \in F$ under this homomorphism by $\overline{w}$. We say that $f : \mathbf{N} \to \mathbf{R}$ is a *Dehn function* for this presentation of G if it satisfies the following condition:
for any freely reduced word w in F such that $\overline{w} = 1$ there are words $r_i \in R$, $p_i \in F(X)$ and $\epsilon_i = \pm 1$ for $i = 1, \ldots, N$ such that

$$w = \prod_{i=1}^{N} p_i r_i^{\epsilon_i} p_i^{-1} \text{ in } F \text{ and } N < f(\ell(w)).$$

It is not hard to see that the existence of a polynomial Dehn function of degree $d \geq 1$ for one finite presentation of the group G implies the existence of a polynomial Dehn function of the same degree d for any other finite presentation. Thus the existence of such a Dehn function is independent of the choice of finite presentation of G. Hence we say that G satisfies a *linear, quadratic, cubic, etc. isoperimetric inequality* if it has a finite presentation with a linear, quadratic, cubic, etc., Dehn function. Similar remarks hold also for Dehn functions which are exponential.

The immediate relevance of these notions is that Cannon et al [CEHPT] have proved that automatic groups, which we have already noted are finitely presented, satisfy a quadratic isoperimetric inequality.

We have alluded previously to the negatively curved groups of M. Gromov [Gr]. There are several equivalent definitions of such groups. One such definition involves so-called *thin triangles*. In order to explain, let G be a finitely generated group and let $\mathcal{A}$ be a finite set of monoid generators of G, $\Gamma_{\mathcal{A}}(G)$ the corresponding Cayley graph. Then Gromov [Gr] has termed a geodesic triangle δ — *thin* if every point on one side of the triangle is no further than δ from at least one point on one of the other two sides, i.e. each side of the triangle is contained in a δ-neighbourhood of the union of the other two sides. The group G is then termed *negatively curved* if there exists a δ such that every geodesic triangle in $\Gamma_{\mathcal{A}}(G)$ is δ-thin. In his fundamental paper [Gr] Gromov proved the remarkable fact that a finitely presented group is negatively curved if and only if it satisfies a linear isopcrimctric inequality.

We have already pointed out that these negatively curved groups are also automatic. They include finitely generated free groups, finite groups, co-compact groups of isometries of n-dimensional hyperbolic space and various classes of small cancellation groups (see Gersten and Short [GS1], [GS2] for the latest word on this subject). In addition it follows readily from B.B. Newman's spelling theorem [N] that one-relator groups with torsion satsify a linear isoperimetric inequality and hence are negatively curved.

13. Amalgamated products of negatively curved groups

Since negatively curved groups are automatic, it makes sense to try to apply Theorem 1 in the case where A and B are negatively curved. To this end, suppose that $\mathcal{A}$ and $\mathcal{B}$ are respectively finite monoid generating sets for A and B and that $L(\mathcal{A})$ is the language of geodesic words over $\mathcal{A} \cup \mathcal{A}^{-1}$ in the Cayley graph of $\Gamma_{\mathcal{A}}(A)$ of A. Then Gromov has proved that $L(\mathcal{A})$ s an automatic language for A over the alphabet $\mathcal{A} \cup \mathcal{A}^{-1}$. The corresponding remark then holds also for B, with $L(B)$ denoting the analogous language of geodesics for B. Then, adopting this notation, the following theorem holds.

THEOREM 3. *Let A and B be negatively curved groups with*

$$G = A \star_C B$$

an amalgamated product of A and B amalgamating a subgroup C that is $L(A)$-regular in A and $L(B)$-regular in B. Then G is asynchronously *automatic. If, in addition, the finite monoid generating sets $\mathcal{A}$ and $\mathcal{B}$ can be so chosen that there exists a constant k' such that for every $g \in C$*

$$| \ell_\mathcal{A}(g) - \ell_\mathcal{B}(g) | \le k',$$

where here $\ell_\mathcal{A}(g)$ denotes the length of a shortest word w over $\mathcal{A} \cup \mathcal{A}^{-1}$ such that $\overline{w} = g$ and similarly for $\ell_\mathcal{B}(g)$, then G is automatic.

Some additional effort is needed to deal with the case where the amalgamated subgroup is cyclic. Here the result takes the following form.

THEOREM 4. *An amalgamated product of two negatively curved groups with a cyclic subgroup amalgamated is automatic.*

Now finitely generated free groups are negatively curved. So Theorem 3 applies also to amalgamated products of finitely generated free groups with a cyclic subgroup amalgamated. Negatively curved groups do not contain free abelian subgroups of rank two [Gr]. On the other hand each torus knot group contains a free abelian group of rank two. So it follows that the free product of two negatively curved groups with a cyclic amalgamation need not be negatively curved. On the other hand Gromov asserts on page 113 of [Gr], §3.3 that if in addition both factors are torsion free and the cyclic subgroup is maximal in both of them, then the resulting product is negatively curved.

14. Amalgamated products of free groups

There is a special case of Theorem 3 that is of independent interest.

THEOREM 5. *Every amalgamated product of two finitely generated free groups with a finitely generated subgroup amalgamated is aysnchronously automatic.*

In order to see that Theorem 5 is indeed a special case of Theorem 3 it suffices to observe that a finitely generated subgroup of a finitely generated free group is a free factor of a subgroup of finite index (M. Hall [H]), or better still, to appeal to the work of Anissimov and Seifert [AS] .

As we pointed out a little earlier, asynchronously automatic groups are finitely presented and have solvable word problem. So it follows, in particular, that the amalgamated products in Theorem 5 all have solvable word problem (this can, of course, be proved directly). On the other hand there exist asynchronously automatic groups with unsolvable conjugacy problem because C.F. Miller III [M] has proved that there are free products of two finitely generated free groups with a finitely generated subgroup amalgamated which have unsolvable conjugacy problem. It is as yet unresolved whether there exist automatic groups with unsolvable conjugacy problem, although the conjugacy problem is solvable for negatively curved groups.

Now Miller [M] has proved that the isomorphism problem for free products of finitely generated free groups with a finitely generated subgroup amalgamated is unsolvable. It follows that the isomorphism problem for asynchronously automatic groups is unsolvable. Again the corresponding problem for automatic groups is still open and considered to be of major importance.

Theorem 5 can be applied also to HNN extensions.

COROLLARY. *Let G be an HNN extension of a finitely generated free group with finitely many stable letters. If the associated subgroups are all finitely generated, then G is asynchronously automatic.*

This corollary follows from Theorem 5 on noting that G is a free factor of an amalgamated product of two finitely generated free groups with a finitely generated subgroup amalgamated and then appealing to the following

PROPOSITION. *A free product of two asynchronously automatic (automatic, negatively curved) groups is asynchronously automatic (automatic, negatively curved) if and only if the factors are asynchronously automatic (automatic, negatively curved).*

As we have already pointed out, the if part of this proposition is due to Cannon et al [CEHPT]. Somewhat surprisingly the proof of the other half is a little delicate. Its counterpart for direct products is unresolved.

15. Amalgamating subgroups of finite index

It is not at all clear just how far one can go with concocting new automatic groups and the like from old ones using amalgamated products. Here is one additional theorem.

THEOREM 6. *A generalised free product*

$$G = A \star_C B$$

*of two asynchronously automatic groups A and B amalgamating a subgroup
C , which is of finite index in both factors, is asynchronously automatic.*

16. An example

It is difficult in general to prove that a given asynchronously automatic
group is not automatic. However we are able to show that the following
HNN extension

$$G = \langle a, b, t, u; tat^{-1} = ab, tbt^{-1} = a, uau^{-1} = ab, ubu^{-1} = a \rangle$$

of the free group of rank two, with two stable letters, is not automatic. As
we pointed out in §14 G is asynchronously automatic. We are however able
to show, by a geometric-topological argument, that G does not satisfy a
quadratic isoperimetric inequality and is consequently not automatic. Now
G is a free factor of an amalgamated product H of two finitely generated
free groups with a finitely generated subgroup amalgamated. So it follows
from the proposition in §14 that there exist amalgamated products of two
finitely generated free groups with a finitely generated subgroup amalgamated which are not automatic. In other words Theorem 5 is best possible.
In a vague sense this example touches on the difficulty in obtaining some
information about the subgroup structure of automatic groups about which
we know very little.

17. Final remarks

Some of the results about amalgamated products that have been described here were obtained while the four authors were participants in the
year long conference on geometric methods in combinatorial group thoery
at MSRI during the fall of 1988. In the past few months Gersten and Short
(jointly) and Shapiro (in some independent work) have developed some new
ideas and perspectives for studying automatic groups. In particular Gersten and Short have extended work of Gilman [Gi] in [GS3]. Shapiro [S] has
studied fundamental groups of finite graphs of automatic groups and has
obtained a version for HNN extensions of our Theorem 1. Many of their
new ideas have been incorporated into the present work.

REFERENCES

[ABCDFLMSS] J. Alonso, T. Brady, D. Cooper, T. Delzant, V. Ferlini, M. Lustig, M. Mihalik, M. Shapiro and H. Short, *Notes on negatively curved groups*, MSRI preprint 1989.

[AS] A. Anissimov and M. Seifert, *Zur algebraischen Charakteritik der durch kontext-freie Sprachen definierten Gruppen*, Elektron. Inform. Verarb. u. Kybernetik **11** (1975), 695–702.

[BGSS] G. Baumslag, S.M. Gersten, M. Shapiro, and H. Short, *Automatic groups and amalgams, preliminary version*, MSRI Preprint No. 09123-90.

[BaS] G. Baumslag and D. Solitar, *Some two generator one-relator non-Hopfian groups*, Bull. Amer. Math. Soc **68**, 199–201.

[C] J.W. Cannon,, *The combinatorial structure of cocompact discrete hyperbolic groups*, Geom. Ded. **16** (1984), 123–148.

[CEHPT] J.W. Cannon, D.B.A. Epstein, D.F. Holt, M.S. Paterson, and W.P.Thurston, *Word processing and group theory*, (preprint).

[CDP] M. Coornaert, T. Delzant, and A. Papadopoulos, *Sur*, preprint, IRMA Strasbourg, 1989.

[ER] D.B.A. Epstein and S. Rees, *Aut programs for automatic structures*, Computer tape from Warwick University.

[G1] S.M. Gersten, *Reducible diagrams and equations*, in "Essays in Group Theory," M.S.R.I. series Vol 8, S.M. Gersten, editor, Springer Verlag, 1987, pp. 15–73.

[Ge] S.M. Gersten, *Dehn functions and l_1−norms of finite presentations*, MSRI preprint 04323-89.

[GS1] S.M. Gersten and H. Short, *Small cancellation theory and automatic groups*, Invent. Math. **102** (1990), 305–334.

[GS2] S.M. Gersten and H. Short, *Small cancellation theory and automatic groups, Part II*, Invent. Math. (to appear).

[GS3] S.M. Gersten and H. Short, *Rational Subgroups of Biautomatic Groups*, To appear, Annals of Mathematics.

[GH] E. Ghys and P. de la Harpe, "Notes sur les groupes hyperboliques de Mikhael Gromov," to appear in Birkhäuser Progress in Math. series, 1990.

[Ge] R. Gilman, *Groups with a rational cross-section*, in "Annals of Mathematics Study," Edited by S.M. Gersten and J. Stallings, Princeton University Press, 1989, pp. 175–183.

[Gr] M. Gromov, *Hyperbolic Groups*, in "Essays in Group Theory,", Springer Ver-lag, Berlin-Heidelberg- New York, 1987.

[H] M. Hall, Jr., *Coset representations in free groups*, Trans. Amer. Math. Soc. **67** (1949), 421–432.

[HU] J.E. Hopcroft and J.D. Ullman, "Formal languages and their relation to automata," Addison-Wesley, 1969.

[L] R.C. Lyndon, *On Dehn's algorithm*, Math. Ann. **166** (1966), 208–228.

[LS] R.C. Lyndon and P.E. Schupp, "Combinatorial group theory," Springer Verlag, 1977.

[MKS] W. Magnus, A. Karrass and D. Solitar, "Combinatorial group theory," Wiley, 1966.

[M] C.F. Miller III, "On group theoretic decision problems and their classification," Annals of Math studies, Princeton University Press, 1971.

[N] B.B. Newman, *Some results in one-relator groups*, Bull Amer Math Soc **74** (1968), 568–571.

[RS] M.O. Rabin and D. Scott, IBM Journal of Research and Development (1959), in "Sequential machines: selected papers," Edited by E.F. Moore, Addison-Wesley,.

[S] M. Shapiro, *Graph products of automatic groups*, Unpublished.

Department of Mathematics, City College of New York, New York NY 10031
Department of Mathematics, University of Utah, Salt Lake City UT 84112
Department of Mathematics, Ohio State University, Columbus OH 43210
Department of Mathematics, City College of New York, New York NY 10031

Dehn Functions and l_1-norms of Finite Presentations

S. M. GERSTEN*

Abstract. An l_1-norm function is introduced as a computational tool to estimate the Dehn function of a finite presentation and settle certain imbedding
questions for finitely presented groups.

1. Introduction

This note represents an initial attempt to decide what are the finitely presented subgroups of an automatic group in the sense of Thurston [CEHPT].
We may start by comparing the situation with Gromov's hyperbolic groups
[Gr]. A hyperbolic group is a finitely generated group whose Cayley graph
in the word metric satisfies the thin triangle condition: there is a positive number $A > 0$ so that for every geodesic triangle XYZ and point
$P \in [X, Y]$ the distance of P from the union of the sides $[X, Z] \cup [Z, Y]$
is bounded by A. An automatic group G is one possessing a finite set of
semigroup generators S and a regular language $\mathcal{L}$ in the free monoid S^*
generated by S so that $\mathcal{L}$ represents every element of G and so that the
question whether any two elements of $\mathcal{L}$ represent elements of G at most
a unit distance apart in the Cayley graph can be decided by a finite state
automaton. For more information on hyperbolic groups and on automatic
groups the reader should consult [BGSS] in this volume; other references
are the original papers [Gr] and [CEHPT] as well as the two fine sets of
notes, [CDP] and [GH].

A result of Gromov's is that a hyperbolic group contains no subgroup
isomorphic to $\mathbf{Z} \oplus \mathbf{Z}$ (see [Gr] Corollary 8.2.C, page 212); this turns out to
be a powerful necessary criterion for a group to be hyperbolic. The same
techniques involving translation lengths along geodesics can be used to
show that no Baumslag-Solitar group $B_{k,l} = G\langle x, y \mid yx^k\bar{y} = x^l \rangle$, $k \cdot l \neq 0$,
imbeds as a subgroup of a hyperbolic group (here and sometimes elsewhere,
the inverse of a generator y is denoted $\bar{y}$). Since it follows from the results
of [GS2] that the groups $B_{k,\pm k}$ are automatic, the natural question to ask
is whether $B_{k,l}$ imbeds as a subgroup of an automatic group if $|k| \neq |l|$.
This question is open in general, but we shall prove the following result.

*This paper is based on research partly supported by the National Science Foundation
grant Nos. DMS 850-5550 and DMS 860-1376

THEOREM A. *If G is the fundamental group of a finite aspherical 2-complex and G is automatic, then $B_{k,l}$ does not imbed as a subgroup of G if $|k| \neq |l|$.*

Our techniques involve calculating a lower bound for the Dehn function δ_X of a finite connected 2-complex X. Here $\delta_X(n)$ is the maximum over all tuples of null homotopic circuits $\mathbf{w}$ of total length at most n of the sum of the minimum number of faces in van Kampen diagrams for the component circuits of $\mathbf{w}$. The Dehn function of a finite presentation is defined to be the Dehn function of the 2-complex canonically associated to the presentation. It turns out that the growth of the function δ_X depends only on the fundamental group of X. Furthermore Gromov [Gr] has shown that $\pi_1(X)$ is hyperbolic iff there exists a constant $A > 0$ such that $\delta_X(n) \leq An$ for all $n > 0$, while Thurston has shown that if $\pi_1(X)$ is automatic, then there exists a constant $A > 0$ such that $\delta_X(n) \leq An^2$ [BGSS].

To estimate the Dehn function we calculate the l_1-norm of the 2-chain $c(\tilde{f})$ determined by a van Kampen diagram $f : D \to X$ for the null homotopic circuit w in $X^{(1)}$; here $\tilde{f} : D \to \tilde{X}$ is a lift of $f : D \to X$ to the universal cover $\tilde{X}$ of X. The l_1-norm of the 2-chain $c(\tilde{f})$ is an invariant $|w|_1$ of w if X is aspherical. The l_1-norm function $\lambda_X(n)$ can then be defined as the supremum of $|\mathbf{w}|_1$ over all tuples of such circuits $\mathbf{w}$ in $X^{(1)}$ of total length at most n, where $|\mathbf{w}|_1$ is the sum of the l_1-norms $|w|_1$ of the component circuits w of $\mathbf{w}$. An analogous definition of λ_X is made in general taking into account the fact that the chain $c(\tilde{f})$ is determined only modulo spherical cycles. The growth of the function λ_X depends only on $\pi_1(X)$ and in general we have $\lambda_X \leq \delta_X$.

We can now refine the statement of theorem A as the consequence of theorems B and C below, which constitute the main results of this note.

THEOREM B. *If X is the 2-complex canonically associated to the Baumslag-Solitar presentation $\langle x, y \mid yx^k\bar{y} = x^l \rangle$ where $|k| \neq |l|$, then $\lambda_X(n)$ grows faster than any polynomial function of n; that is, there are no constants $A, d > 0$ with $\lambda_X(n) \leq An^d$ for all $n > 0$.*

THEOREM C. *Let $f : X \to Y$ be a map of finite aspherical 2-complexes such that the induced homomorphism on fundamental groups is injective. Then there exist positive constants $A, B,$ and C such that $\lambda_X(n) \leq A\lambda_Y(Bn) + Cn$ for all $n > 0$.*

By the results of [BGSS] the Baumslag-Solitar groups $B_{k,l}$ are all asynchronously automatic and hence their Dehn functions' growth is bounded by a simple exponential, where a simple exponential function of n is of the form A^n for some constant $A > 0$. In §6 we give a number of examples of the use of our results and we produce some examples of even more rapidly growing Dehn functions. In particular we make use of Dehn functions in section 6 to settle negatively a question raised by Baumslag, whether a 1-relator group of deficiency at least 2 is an amalgam of finitely generated free groups amalgamating finitely generated subgroups; our example $\langle x, y, z \mid x^{x^y} = x^2 \rangle$, where $x^y = yx\bar{y}$, has a Dehn function which grows faster than any simple exponential.

In section 5 we show that a split extension $\mathbf{Z}^2 \rtimes_\phi \mathbf{Z}$ is automatic iff the automorphism $\phi \in \mathrm{Gl}_2(\mathbf{Z})$ is of finite order. In particular this establishes the fact that the 3-dimensional integral Heisenberg group is not automatic; more precisely, we show that its Dehn function is bounded by a cubic polynomial and this bound is optimal. We also show that the infinite family of words used to establish the cubic bound on the Dehn function in the 3-dimensional Heisenberg group has a quadratic bound on areas, when considered in the 5-dimensional integral Heisenberg group.

This article is based on an address given in the Workshop on Algorithmic Problems held at MSRI in January, 1989. We should like to thank the referee for his careful reading of this paper and for his suggestions, which we have adopted to aid the exposition.

2. The Function δ_X

We shall consider combinatorial 2-complexes and combinatorial maps [G]. Recall that a combinatorial map is one whose restriction to each open cell of each dimension is a homeomorphism onto its image. A combinatorial 2-complex is one whose attaching maps are combinatorial maps. We shall use freely the terminology of van Kampen diagrams [LS]. Recall that a van Kampen diagram is a combinatorial map whose domain is a contractible plane point set. In addition we shall need at one point to consider the quotient complex of a combinatorial 2-complex by a subcomplex. There is a canonical way of giving the quotient the structure of a combinatorial 2-complex that is discussed in [G].

A presentation $\mathcal{P}$ gives rise to a combinatorial 2-complex $K(\mathcal{P})$ whose 0-skeleton consists of a single point, whose edges are generators and whose 2-cells are attached by maps that spell out the relators. Presentations and such 2-complexes are essentially the same thing, where the relators are determined up to cyclic permutation of letters and order of relators.

DEFINITION 2.1: Let X be a finite connected combinatorial 2-complex. Let $\mathbf{w} = (w_1, w_2, \ldots, w_r)$ be an ordered set (for variable r) of edge circuits in the 1-skeleton each of which is null homotopic in X. We define the length $\ell(\mathbf{w}) = \sum_{i=1}^{r} \ell(w_i)$, where $\ell(w_i)$ is the length of w_i. We define $\text{Area}_X(w_i)$ to be the minimal number of faces (=2-cells) in a van Kampen diagram for w_i and we set $\text{Area}_X(\mathbf{w}) = \sum_i \text{Area}_X(w_i)$. We define the Dehn function δ_X (or *isoperimetric function*) on the natural numbers by setting $\delta_X(n) = \max_{\ell(\mathbf{w}) \leq n} \text{Area}_X(\mathbf{w})$.

The motivation for this definition comes from differential geometry, where we span a null homotopic loop w by a minimal surface and calculate the area of this surface. The fact that we have to consider families of null homotopic loops here is a technicality to make our definitions homotopy invariant.

If $\mathcal{P}$ is a finite presentation, we set $\delta_{\mathcal{P}} = \delta_{K(\mathcal{P})}$.

The interest in the Dehn function of a finite presentation $\mathcal{P}$ is explained by the next result. Here $G(\mathcal{P}) = \pi_1(K(\mathcal{P}))$ is the group of the presentation.

PROPOSITION 2.2. *If $\mathcal{P}$ is a finite presentation, then the group $G(\mathcal{P})$ has a solvable word problem iff $\delta_{\mathcal{P}}$ is recursive.*

PROOF: Assume first that $\delta_{\mathcal{P}}$ is recursive and a word w in the generators is given. One can then calculate $\delta_{\mathcal{P}}(\ell(w))$, where w is identified with the 1-tuple (w), to obtain an effective bound on the number of relators and their inverses in an expression for w as a product of conjugates of them. We can also obtain an effective bound on the lengths of the words performing the conjugations on the relators as follows. If K is a minimal van Kampen diagram for w, we let $\text{Area}(K)$ be the number of faces and let $\text{girth}(K)$ be the length of the longest edge path in $K^{(1)}$ which does not contain a circuit. The lengths of words performing the conjugations are then bounded by $\text{girth}(K)$ and we have the estimate

$$\text{girth}(K) \leq L \, \text{Area}(K) + \ell(w),$$

where L is the length of the longest relator in $\mathcal{P}$. Since $\text{Area}(K)$ is effectively estimated by $\delta_{\mathcal{P}}(\ell(w))$, this gives an effective bound on the lengths of the conjugating words. It follows that the words of length at most n which represent 1 in $G(\mathcal{P})$ can be effectively listed and one can hence check whether or not a given word w lies on this list. Thus $G(\mathcal{P})$ has a solvable word problem if $\delta_{\mathcal{P}}$ is recursive.

Conversely assume that $G(\mathcal{P})$ has a solvable word problem. Then it is decidable whether each ordered r-tuple $\mathbf{w} = (w_1, w_2, \ldots, w_r)$ of words is such that each word w_i represents 1. In particular, for such tuples $\mathbf{w}$, one can calculate effectively an expression for each component as a product of conjugates of relators and inverses of relators. Thus one can also calculate a minimal such expression, involving the fewest relators and inverse relators (making use of the girth estimate in the preceding paragraph). Thus Area$_\mathcal{P}(\mathbf{w})$ is calculable for such tuples $\mathbf{w}$ and consequently the maximal area of such tuples $\mathbf{w}$ with $\ell(\mathbf{w}) \leq n$ is effectively calculable. That is, $\delta_\mathcal{P}$ is effectively calculable.

DEFINITION 2.3: Suppose that f and g are two functions defined on $\mathbf{N}$ taking non negative real values. We write $f \sim g$ if there are positive constants A, A', B, B', C and C' such that $f(n) \leq Ag(Bn) + Cn$ and $g(n) \leq A'f(B'n) + C'n$ for all $n \in \mathbf{N}$. This relation is clearly an equivalence relation, and we say that two equivalent functions have the same 'growth'.

PROPOSITION 2.4. *If $\mathcal{P}$ and $\mathcal{Q}$ are finite presentations for isomorphic groups, then $\delta_\mathcal{P} \sim \delta_\mathcal{Q}$.*

PROOF: We shall show that the Dehn functions for Tietze equivalent presentations have the same growth. Suppose that $\mathcal{P} = \langle \mathcal{X} \mid \mathcal{R} \rangle$ and suppose that $\mathcal{Q}$ is obtained from $\mathcal{P}$ by adjoining one additional relator R, where R is a consequence of $\mathcal{R}$. If all components of the tuple $\mathbf{w}$ represent 1, then we clearly have Area$_\mathcal{P}(\mathbf{w}) \geq$ Area$_\mathcal{Q}(\mathbf{w})$, where $\mathbf{w}$ is a tuple of words in the generators $\mathcal{X}^{\pm 1}$. On the other hand, the word R is the boundary label of a van Kampen diagram in $\mathcal{P}$. If A is the number of faces in one such $\mathcal{P}$ diagram for R, then we see by subdividing the faces labelled R in $\mathcal{Q}$ diagrams that Area$_\mathcal{P}(\mathbf{w}) \leq A \cdot$ Area$_\mathcal{Q}(\mathbf{w})$. It follows from these observations that $\delta_\mathcal{P} \sim \delta_\mathcal{Q}$.

Now suppose that $\mathcal{Q}$ is obtained from $\mathcal{P}$ by adjoining one free generator $t \notin \mathcal{X}$ and adjoining one relator tr^{-1} with r a word of length L in the generators $\mathcal{X}^{\pm 1}$. If $\mathbf{w}$ is a tuple of words in $\mathcal{X}$ each representing 1 in $G(\mathcal{P})$, then Area$_\mathcal{P}(\mathbf{w}) =$ Area$_\mathcal{Q}(\mathbf{w})$. This is true since a face with label tr^{-1} can never have an edge labelled t as an interior edge in a minimal diagram for $\mathbf{w}$, for otherwise that diagram would not be reduced.

Suppose next that $\mathbf{u}$ is a tuple of words in the generators of $\mathcal{Q}$ representing 1 and $n = \ell(\mathbf{w})$. We have already seen that an edge labelled t cannot occur an an interior edge of a minimal diagram for a component of $\mathbf{u}$. The only occurrences of t in such a diagram occur on the boundary, and there are at most n of them in all the components of $\mathbf{u}$. Thus there are at most a total of n faces with label tr^{-1} in all the minimal diagrams for the components. Next replace each occurrence of t in a component word of $\mathbf{u}$ by the word r (without doing any cancellations) to obtain a tuple $\mathbf{w}$ in $\mathcal{X}^{\pm 1}$. Observe that $\ell(\mathbf{w}) \leq nL$ and all components of $\mathbf{w}$ represent 1. Given minimal $\mathcal{P}$ diagrams for the components of $\mathbf{w}$ we obtain $\mathcal{Q}$ diagrams for the components of $\mathbf{u}$ by attaching at most n faces in total with label tr^{-1} along the boundaries. Hence $\text{Area}_{\mathcal{Q}}(\mathbf{u}) \leq \text{Area}_{\mathcal{P}}(\mathbf{w}) + n$ and consequently $\delta_{\mathcal{Q}}(n) \leq \delta_{\mathcal{P}}(Ln) + n$. Taken with the result of the preceding paragraph, this yields $\delta_{\mathcal{P}} \sim \delta_{\mathcal{Q}}$.

It follows from these calculations and from Tietze's theorem that if $\mathcal{P}$ and $\mathcal{Q}$ are two finite presentations for isomorphic groups, then $\delta_{\mathcal{P}} \sim \delta_{\mathcal{Q}}$. This completes the proof of Proposition 2.4.

We want to extend this result about presentations to 2-complexes. The crucial step in the argument turns out to be an invariance under free product with an infinite cycle. First we observe

LEMMA 2.5. *For any finite connected 2-complex X we have $\delta_X(m+n) \geq \delta_X(m) + \delta_X(n)$.*

PROOF: Let $\mathbf{u} = (u_1, u_1, \ldots, u_r)$ and $\mathbf{v} = (v_1, v_2, \ldots, v_s)$ be tuples of null homotopic edge circuits in $X^{(1)}$ with $\ell(\mathbf{u}) \leq m$ and $\ell(\mathbf{v}) \leq n$. We may assume that $\text{Area}_X(\mathbf{u})$ and $\text{Area}_X(\mathbf{v})$ are maximal among all tuples with lengths in these ranges. Let $\mathbf{w} = (u_1, u_2, \ldots, u_r, v_1, \ldots, v_s)$ Then $\text{Area}_X(\mathbf{w}) = \text{Area}_X(\mathbf{u}) + \text{Area}_X(\mathbf{v}) = \delta_X(m) + \delta_X(n)$ and $\ell(\mathbf{w}) \leq m + n$. From this follows the desired inequality.

PROPOSITION 2.6. *Let $\mathcal{P}$ be a finite presentation and let $\mathcal{Q}$ be obtained from $\mathcal{P}$ by adjoining one new free generator t and leaving the relators unchanged. Then $\delta_{\mathcal{P}} = \delta_{\mathcal{Q}}$.*

PROOF: Since an edge labelled t cannot appear in the interior of a reduced diagram, it is clear that $\delta_{\mathcal{P}} \leq \delta_{\mathcal{Q}}$. We shall establish the opposite inequality $\delta_{\mathcal{Q}}(n) \leq \delta_{\mathcal{P}}(n)$ by induction on n. The induction starts trivially with $n = 0$. In the inductive step, assume the inequality is established for numbers less than n, $n > 0$, and let $\mathbf{w}$ be a tuple of words representing 1 in $\mathcal{Q}$ with $\ell(\mathbf{w}) = n$. If none of the component words of $\mathbf{w}$ contains t, then $\text{Area}_{\mathcal{Q}}(\mathbf{w}) = \text{Area}_{\mathcal{P}}(\mathbf{w})$. If the component word w_i involves t, then a minimal van Kampen diagram K for w_i cannot be a disc (since an edge labelled t is not in the boundary of any face). We replace w_i with the a set of words obtained by reading the boundary labels of the disc components of K; the disc subdiagram of K for each of these latter words must be minimal, for otherwise K would not be a minimal diagram for w_i. Thus the area is unchanged by replacing w_i by this collection, but the length is reduced by at least one. We replace $\mathbf{w}$ by a new tuple whose components in some ordering are the old components w_j for $j \neq i$ and an ordering of the collection of words with which we replaced w_i. This does not change the area but decreases the length, so the induction hypothesis can apply. Observe that this last step uses the fact that $\delta_{\mathcal{P}}$ is monotone increasing; this fact follows from Lemma 2.5.

THEOREM 2.7. *If X and X' are finite connected 2-complexes with isomorphic fundamental groups, then $\delta_X \sim \delta_{X'}$.*

PROOF: We shall establish that $\delta_X \sim \delta_{\mathcal{P}}$, where $\mathcal{P}$ is a finite presentation for $\pi_1(X, x)$. This reduces the Theorem to the earlier result, Proposition 2.4, about presentations.

Let T be a maximal tree in the 1-skeleton of X and let $p : X \to Y := X/T$ be the quotient map. We recall here that Y is given the structure of a combinatorial 2-complex by the method of [G]. Let $q : X \to X/X^{(0)} :=$ Z be the quotient map. Note that p is a homotopy equivalence whereas q imbeds the fundamental group of X as a free factor of $\pi_1(Z)$, where the complementary factor is freely generated by a set of oriented edges of T. Since Y and Z are 1-vertex 2-complexes, they can be considered as presentations, and we shall do so. Let D denote the diameter of the tree T.

Suppose now that $\mathbf{w}$ is a tuple of null homotopic edge circuits in Y with $\ell(\mathbf{w}) = n$. We can lift each component and hence lift $\mathbf{w}$ to a tuple $\mathbf{u}$ of null homotopic edge circuits in X, where $\ell(\mathbf{u}) \leq (D+1)\ell(\mathbf{w})$. By considering minimal diagrams for the components of $\mathbf{u}$ we see that $\mathrm{Area}_Y(\mathbf{w}) \leq \mathrm{Area}_X(\mathbf{u}) \leq \delta_X((D+1)n)$. Hence $\delta_Y(n) \leq \delta_X((D+1)n)$.

Suppose now that $\mathbf{v}$ is a tuple of null homotopic edge circuits in X. Then we have $\mathrm{Area}_X(\mathbf{v}) = \mathrm{Area}_Z(q(\mathbf{v}))$, since identifying vertices does not change the class of diagrams. Here $q(\mathbf{v})$ is the image tuple in Z. But this means that $\delta_X \leq \delta_Z$. Recalling that Z is essentially a presentation, we see that $\delta_Y \sim \delta_Z$ by applying Propositions 2.6 and 2.4. It follows that $\delta_X \sim \delta_Y$. Since Y can be considered a presentation, this completes the proof of the Theorem.

REMARK: If we were considering only presentations, we could have defined a simpler notion of the Dehn function, by considering only words representing 1 instead of tuples of words. This more complicated definition was needed to assure that Proposition 2.6 holds.

The next two assertions, 2.8 and 2.9, represent substantial theorems.

THEOREM 2.8 (GROMOV). *If $\mathcal{P}$ is a finite presentation, then $G(\mathcal{P})$ is hyperbolic iff there is a constant $A > 0$ such that for all $n > 0$ one has $\delta_{\mathcal{P}}(n) \leq An$.*

An elementary proof of the difficult 'if' direction due to H. Short can be found in the appendix of [GS1].

THEOREM 2.9 (THURSTON). *If $\mathcal{P}$ is a finite presentation such that the group $G(\mathcal{P})$ is automatic, then there is a constant $A > 0$ such that for all $n > 0$ one has $\delta_{\mathcal{P}}(n) \leq An^2$, (the so called 'quadratic isoperimetric inequality') .*

It is proved in [CEHPT] that if $G(\mathcal{P})$ is automatic, then there is a constant $A > 0$ such that for all words w representing 1 one has $\mathrm{Area}_{\mathcal{P}}(w) \leq A\ell(w)^2$. Using the convexity of the squaring function $x \to x^2$, it follows that $\mathrm{Area}_{\mathcal{P}}(\mathbf{w}) \leq A\ell(\mathbf{w})^2$. Hence $\delta_{\mathcal{P}}(n) \leq An^2$.

The converse to Theorem 2.9 is false. Thurston [Th] asserts that the 5-dimensional integral Heisenberg group $G(\langle x_1, x_2, y_1, y_2, z \mid [x_1, x_2] = z = [y_1, y_2]; [x_i, y_j] = 1 \text{ for all } i, j; z \text{ central } \rangle)$ satisfies the quadratic isoperimetric inequality but is not automatic.

THEOREM 2.10. *All finite small cancellation presentations $\mathcal{P}$ and all finitely presented subgroups of $G(\mathcal{P})$ satisfy the quadratic isoperimetric inequality.*

A small cancellation presentation $\mathcal{P}$ is one having the property that for any reduced van Kampen diagram D in $\mathcal{P}$, the associated diagram $\bar{D}$ is of *non positive curvature*; here $\bar{D}$ is obtained by removing interior vertices of valence two from D and declaring each face F of $\bar{D}$ to be a regular Euclidean n-gon of unit side, if F has n sides. To say that $\bar{D}$ is of non positive curvature is to say that the angle sum at each interior vertex of $\bar{D}$ is at least 2π. The quadratic isoperimetric inequality here follows from a fundamental result of Reshetnyak's [Re]. The result for finitely presented subgroups follows from a covering space argument.

REMARK: It is an open question whether a group possessing a finite small cancellation presentation is automatic. It is shown in [GS1] and [GS2] that all groups possessing finite $C(p) - T(q)$ presentations are automatic, if $(p, q) = (3, 6)$, $(4, 4)$, or $(6, 3)$. Here the $C(p)$ condition means that each face of $\bar{D}$ has at least p sides and the $T(q)$ condition means that each interior vertex of $\bar{D}$ has valence at least q. One may also ask whether the fundamental group of a compact space of non positive curvature, in the general sense Gromov has defined it [Gr] in terms of the CAT(0) inequality, is automatic.

We end this section with a useful result that demonstrates why it is necessary to formulate the notion of Dehn function for 2-complexes and not just for finite presentations.

PROPOSITION 2.11. *Let G be a finitely presented group and let $H < G$ be a subgroup of finite index. If $\mathcal{P}$ and $\mathcal{Q}$ are finite presentations for G and H respectively, then $\delta_{\mathcal{P}} \sim \delta_{\mathcal{Q}}$.*

PROOF: Let X be the 2-complex canonically associated to $\mathcal{P}$ and let $p : Y \rightarrow X$ be the covering space corresponding to $H < G = \pi_1(X)$. If w is an edge circuit in $X^{(1)}$ which is null homotopic in X, then w is covered by an edge circuit $\tilde{w}$ in $Y^{(1)}$ and $\mathrm{Area}_X(w) = \mathrm{Area}_Y(\tilde{w})$; both of these assertions follow from the covering homotopy theorem, since a van Kampen diagram is simply connected. From this observation one deduces easily that $\delta_X \sim \delta_Y$. The result then follows from Theorem 2.7.

3. The Function λ_X

If A is a free abelian group with given basis $\{e_i \mid i \in I\}$ and if $a = \sum_{i \in I} n_i e_i \in A$, where $n_i = 0$ for all but finitely many indices $i \in I$, we set $|a|_1 = \sum_i |n_i|$, the l_1-norm of a. If Y is a 2-complex, then the group of 2-chains with integral coefficients $C_2(Y)$ has a preferred basis of 2-cells, so we have the l_1-norm defined.

Suppose now that X is a finite connnected 2-complex and w is a circuit in $X^{(1)}$ with w null homotopic in X. If we choose a van Kampen diagram $f : D \rightarrow X$, we can lift f to the universal cover $\tilde{X}$ to get $\tilde{f} : D \rightarrow \tilde{X}$ with $p \cdot \tilde{f} = f$; here $p : \tilde{X} \rightarrow X$ is the universal cover of X.

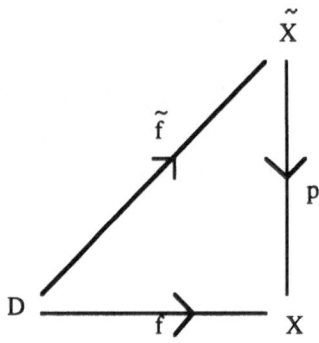

Figure 1

If we choose base points we get a chain $c(\tilde{f}) \in C_2(\tilde{X})$. The l_1-norm $|c(\tilde{f})|_1$ is independent of the choice of base points and of the lift $\tilde{f}$. In fact, if $g : \tilde{X} \to \tilde{X}$ is a deck transformation, then $|c(g \cdot \tilde{f})|_1 = |c(\tilde{f})|_1$.

A more serious matter is the dependence on f. If $f' : D' \to X$ is another van Kampen diagram for w, we can glue f and f' together to get a map $h : S^2 \to X$ with lift $\tilde{h} : S^2 \to \tilde{X}$. Then $c(\tilde{h}) = c(\tilde{f}) - c(\tilde{f'}) \in Z_2(\tilde{X})$, where $Z_2(\tilde{X})$ is the cycle group in dimension 2. Recall that by the Hurewicz theorem $Z_2(\tilde{X}) = \pi_2(X)$. We set

$$(3.1) \qquad |w|_1 = \inf_{z \in Z_2(\tilde{X})} |c(\tilde{f}) + z|_1.$$

We note the following result which follows easily from the definitions.

LEMMA 3.2. *For each null homotopic edge circuit w in X and lift $\tilde{f}$ of the van Kampen diagram f for w we have $|w|_1 \leq |c(\tilde{f})|_1$. In addition, $|c(\tilde{f})|_1$ is at most as large as the number of faces in the domain of f.*

We note the next result which shows how to calculate $|w|_1$ in one case.

LEMMA 3.3. *If $\pi_2(X) = 0$, then $|w|_1 = |c(\tilde{f})|_1$ for any van Kampen diagram for w.*

PROOF: Since $\pi_2(X) = Z_2(\tilde{X}) = 0$, the indeterminacy z in the definition 3.1 of $|w|_1$ is zero.

Suppose now that $\mathbf{w} = (w_1, w_2, \ldots, w_r)$ is an r-tuple of edge circuits in $X^{(1)}$, each of which is null homotopic in X. We define $|\mathbf{w}|_1 = \sum_i |w_i|_1$ and we define $\lambda_X(n)$ as the maximum value of $|\mathbf{w}|_1$ over all tuples $\mathbf{w}$ satisfying $\ell(\mathbf{w}) \leq n$.

The next result is immediate from Lemma 3.2.

LEMMA 3.4. *For all finite connected 2-complexes X we have $\lambda_X \leq \delta_X$.*

THEOREM 3.5. *If X and Y are finite connected 2-complexes with isomorphic fundamental groups, then $\lambda_X \sim \lambda_Y$.*

This result is proved by following the same steps leading to the proof of Theorem 2.7. One first defines $\lambda_\mathcal{P} = \lambda_{K(\mathcal{P})}$, where $\mathcal{P}$ is a finite presentation. The 'growth' of $\lambda_\mathcal{P}$ is shown to be invariant under Tietze transformation. One proves the analog of Proposition 2.6 and deduces from it that $\lambda_X \sim \lambda_\mathcal{P}$, where $\mathcal{P}$ is a finite presentation for $\pi_1(X, x)$. Since the arguments are copies of those in §2, we omit them.

REMARK 3.6: Suppose that the Hurewicz map $\pi_2(X) \to H_2(X)$ is zero (such a 2-complex X is called *Cockcroft*). Then the projection map p : $\tilde{X} \to X$ kills spherical cycles. Thus if $c(f) \in C_2(X)$ is the chain of the van Kampen f for the null homotopic edge circuit w, we have $|c(f)|_1 \leq |w|_1$. Hence for a Cockcroft 2-complex X one can estimate the l_1-norm $|w|_1$ from below in X without appealing to $\tilde{X}$. The same remark applies to tuples $\mathbf{w}$ of null homotopic circuits. The next result gives us examples of Cockcroft 2-complexes, which we shall make use of in §5.

PROPOSITION 3.7. *Suppose that M is a CW-complex which is a closed, orientable, aspherical 3-manifold and suppose that M possesses exactly one 3-cell. Then the 2-skeleton $M^{(2)}$ of M is a Cockcroft 2-complex.*

PROOF: From the exact homology sequence for the pair $(M, M^{(2)})$, we see that the map $H_2(M^{(2)}) \to H_2(M)$ is injective. The composition of maps $\pi_2(M^{(2)}) \to \pi_2(M) \to H_2(M)$ is zero since M is aspherical. We deduce by a diagram chase that the Hurewicz map $\pi_2(M^{(2)}) \to H_2(M^{(2)})$ is zero. Hence $M^{(2)}$ is Cockcroft.

PROPOSITION 3.8. *Suppose that $\mathcal{P}$ is a finite presentation such that there exists words w_n in the generators representing 1 in $G(\mathcal{P})$ with $\ell(w_n) \to \infty$ as $n \to \infty$, and such that for no positive constant A is it the case that $|w_n|_1 \leq A\ell(w_n)^2$ for all $n > 0$. Then $G(\mathcal{P})$ is not automatic.*

PROOF: If $G(\mathcal{P})$ were automatic, then by Theorem 2.9 there exists $A > 0$ such that $\delta_\mathcal{P}(n) \leq An^2$ for all $n > 0$. But then we would have $|w_n|_1 \leq \lambda_\mathcal{P}(\ell(w_n)) \leq \delta_\mathcal{P}(\ell(w_n)) \leq A\ell(w_n)^2$ for all $n > 0$, which is contrary to hypothesis.

4. Proof of Theorem B

In this section we fix $\mathcal{P} = \langle x, y \mid yx^k\bar{y} = x^l \rangle$ with $1 \leq k < l$. We shall construct words w_n in the generators of $\mathcal{P}$ such that $|w_n|_1$ increases more rapidly than any polynomial in $\ell(w_n)$; that is, for no constants $A, d > 0$ does one have $|w_n|_1 \leq A \cdot \ell(w_n)^d$ for all $n > 0$. This will establish Theorem B of §1. In addition it then follows from Proposition 3.8 that $G(\mathcal{P})$ is not automatic. The stronger assertion that $G(\mathcal{P})$ cannot even be imbedded as a subgroup of the fundamental group of a finite aspherical 2-complex follows from Theorem C, which will be proved in the next section. Similar considerations apply to the other Baumslag-Solitar presentations in Theorem B when $|k| \neq |l|$; the modifications of the arguments given here are straightforward and are left to the reader.

Observe that $X = K(\mathcal{P})$ is aspherical, by Lyndon's theorem [LS] Proposition 11.1, page 161. Hence by Lemma 3.3, $|w|_1 = |c(\tilde{f})|_1$ where $f : D \to X$ is any van Kampen diagram for w and $\tilde{f}$ is a lift to $\tilde{X}$.

We define inductively two sequences of non negative integers, a_i for $i \geq 1$ and b_j for $j \geq 0$, satisfying $a_1 = k$, $b_0 = 0$, and $k|a_i$, $0 \leq b_i \leq k-1$ for all i. If a_i and b_i have been defined with these properties for all $i \leq r$, we let b_{r+1} be the least natural number with $\frac{l}{k}a_r + b_{r+1}$ divisible by k, and we set $a_{r+1} = \frac{l}{k}a_r + b_{r+1}$. One sees immediately that the induction continues. The desired words w_n are defined by

$$w_n = [y^n x^k \bar{y} x^{b_2} \bar{y} x^{b_3} \ldots \bar{y} x^{b_n} \bar{y}, x]$$

where $[a, b] = ab\bar{a}\bar{b}$. Observe that since $0 \leq b_i < k$ we have $\ell(w_n) \leq 2(k+2)n + 2$, which is bounded by a linear function of n. In addition w_n represents 1 in $G(\mathcal{P})$. To see this observe that $u_n = y^n x^k \bar{y} x^{b_2} \ldots \bar{y} x^{b_n} \bar{y}$ represents $x^{\frac{l}{k}a_n}$ and a van Kampen diagram $\mathcal{D}_n$ for w_n is obtained by gluing two copies of a diagram for $v_n = u_n \cdot \bar{x}^{\frac{l}{k}a_n}$ together along a segment of the boundary labelled $x^{\frac{l}{k}a_n - 1}$. See Figure 2 below.

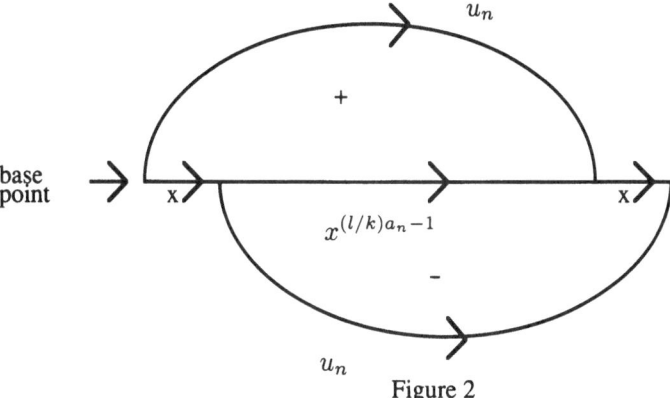

Figure 2

We show schematically a van Kampen diagram $\mathcal{E}_n$ for v_n below in Figure 3, drawn for $n = 4$.

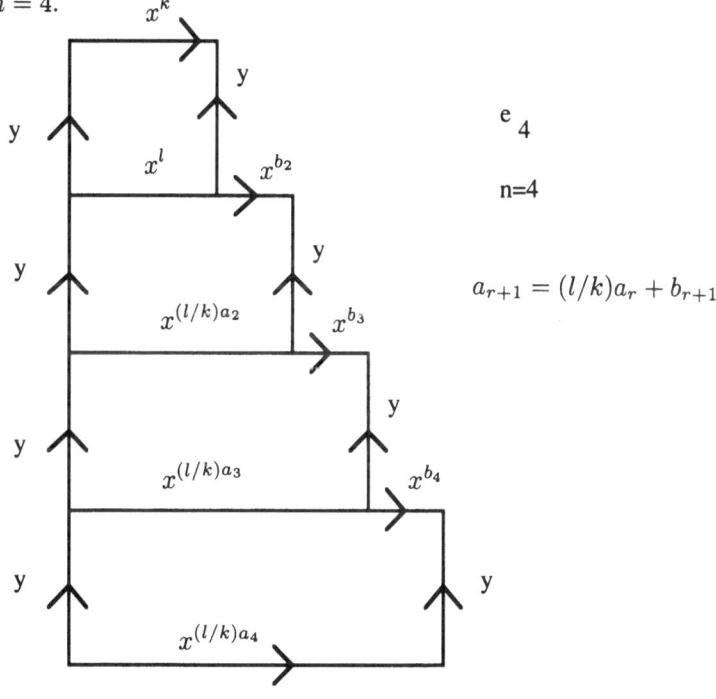

Figure 3

The number of faces of $\mathcal{E}_n$ is $\frac{1}{k}(a_1 + a_2 + \cdots + a_n) \geq \frac{1}{k}(1 + \frac{l}{k} + \cdots + (\frac{l}{k})^{n-1}) = \frac{1}{k} \cdot \frac{l^n - k^n}{l - k}$. Thus if $\mathcal{D}_n$, the diagram constructed for w_n, has F_n faces, then $F_n \geq \frac{2}{k}\frac{l^n - k^n}{l - k}$. Since this number increases exponentially with

n whereas $\ell(w_n)$ increases at most linearly, it follows that F_n increases more rapidly than any polynomial in $\ell(w_n)$. If we can show that there is no cancellation in the chain associated to a lift of $\mathcal{D}_n$ to $\tilde{X}$ between any of the lifted faces, then it will follow that $|w_n|_1$ increases faster than any polynomial in $\ell(w_n)$. This will establish Theorem B.

We chose the base point in the 2-cell α of X as shown in Figure 4.

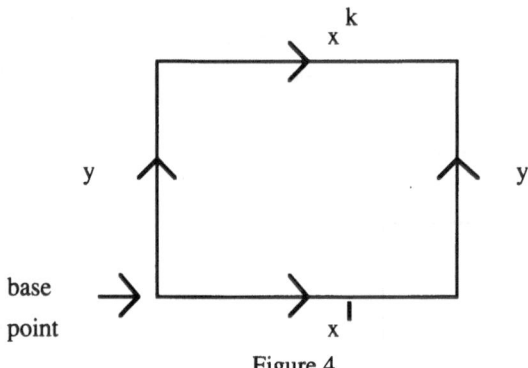

Figure 4

The base point in the domain of $\mathcal{D}_n$ is shown in Figure 2. If $\tilde{\alpha}$ is a lift of α to $\tilde{X}$, then $c(\tilde{\mathcal{D}}_n) = (1 - x) \cdot c(\tilde{\mathcal{E}}_n)$ where the chains have coefficients in the integral group ring $\mathbf{Z}G$ with $G = \pi_1(X)$. We have

$$c(\tilde{\mathcal{E}}_n) = \{(1 + x^l + x^{2l} + \cdots) + y(1 + x^l + x^{2l} + \cdots)$$

$$+ y^2(1 + x^l + \cdots) + \cdots + y^{n-1}\} \cdot \tilde{\alpha}.$$

Thus we must show there is no cancellation among the terms of the following expression in $\mathbf{Z}G$:

$$(4.1) \quad (1 - x)\{(1 + x^l + x^{2l} + \cdots) + y(1 + x^l + x^{2l} + \cdots) + \cdots + y^{n-1}\}$$

It is convenient to work in the group ring of a homomorphic image H of G. We map G into 2x2 upper triangular matrices over $\mathbf{Q}$ by

$$x \mapsto \xi = \begin{pmatrix} 1 & 1 \\ 0 & 1 \end{pmatrix}$$

$$y \mapsto \eta = \begin{pmatrix} \frac{l}{k} & 0 \\ 0 & 1 \end{pmatrix}.$$

One checks the relation $\eta\xi^k\bar\eta = \xi^l$ is satisfied. In the homomorphic image fractional powers of ξ make sense; for example, $\xi^{\frac{l}{k}}$ means the matrix $\begin{pmatrix} 1 & \frac{l}{k} \\ 0 & 1 \end{pmatrix}$. These fractional powers of ξ represent distinct elements of H, so are independent over $\mathbb{Z}$. The image of the expression (4.1) in $\mathbb{Z}H$ is hence

$$(4.2) \qquad (1-\xi)\{(1+\xi^l+\xi^{2l}+\cdots)+(1+\xi^{\frac{l}{k}l}+\xi^{2\frac{l}{k}l}+\cdots)\eta$$

$$+(1+\xi^{(\frac{l}{k})^2l}+\cdots)\eta^2+\cdots+\eta^{n-1}\}.$$

Since terms with different powers of η in the last expression cannot cancel, the only potential cancellation is in the coefficient series of a single power of η. But the coefficients of η^r are of the form $\xi^{(\frac{l}{k})^r il}$ and $-\xi^{(\frac{l}{k})^r il+1}$, where $i \geq 0$. We need

LEMMA 4.3. For all $r, i, j \geq 0$ one has $(\frac{l}{k})^r il \neq (\frac{l}{k})^r jl + 1$.

PROOF: Otherwise we would have $l^{r+1}(i-j) = k^r$, which contradicts the initial assumption $l > k \geq 1$.

It follows that there is no cancellation in (4.2) and consequently none in (4.1). Thus $|w_n|_1 = F_n$, which we saw increases faster than any polynomial in $\ell(w_n)$. This completes the proof of Theorem B.

REMARK 4.4: In the special case $\langle x, y \mid yx\bar y = x^2 \rangle$ all the numbers $b_i = 0$ and the diagram $\mathcal{D}_n$ has the form illustrated below in Figure 5 for $n = 3$.

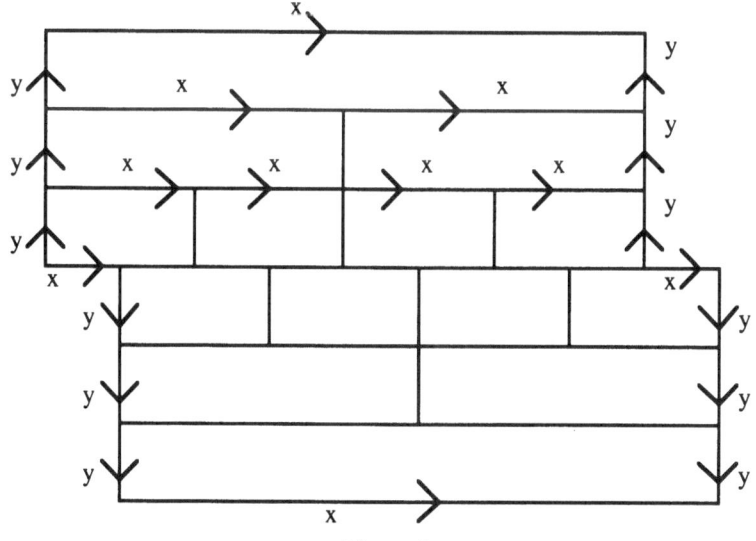

Figure 5

We shall now indicate a geometrical proof of Theorem B based on a different principle. The starting point is the disc diagrams $\tilde{\mathcal{D}}_n$ in $\tilde{X}$. To show that the l_1-norm of the chain $c(\tilde{\mathcal{D}}_n)$ in $C_2(\tilde{X})$ is equal to the number of faces in the domain, it suffices to show that the mapping $\tilde{f}_n$ from the domain of $\tilde{\mathcal{D}}_n$ to $\tilde{X}$ is injective, for if there is cancellation in the chain, then the corresponding faces must have the same image in $\tilde{X}$.

DEFINITION 4.5: Let Y be a piecewise Euclidean 2 complex of non positive curvature. Recall that this means each cell of Y has the metric of an convex polygon in $\mathbf{R}^2$ and the metrics agree on overlapping edges (there may not exist any metric on Y inducing the given metric on the cells). The condition of non positive curvature means that every non trivial circuit without backtracking in the link of each vertex of Y has total angular measure at least 2π. Let $f : D \to Y$ be a reduced disc diagram in Y (so D is a topological disc) and give D the pull back piecwise Euclidean structure; thus D is of non positive curvature. It follows from results of Aleksandrov [A] that (1) the path metric on D agrees with the given metric on each cell of D and (2) D possesses unique geodesic segments. The geodesics here are defined in a purely local manner, the condition being that all paths in the link connecting the incoming ray to the outgoing ray have angular measure at least π.

LEMMA 4.6. If $f : D \to Y$ is a reduced disc diagram and Y is a piecewise Euclidean 2-complex of non positive curvature and if the angle sum at each interior vertex of D is precisely 2π, then f maps geodesic segments γ of D, which meet ∂D only at end points of γ, to geodesic segments of Y.

PROOF: We need only check the angle condition at a vertex $f(P)$ of $f(D)$, where P is a vertex lying on the interior of γ. But if some path in $\text{Link}_Y(f(P))$ connecting the incoming ray to the outgoing ray had angular measure less than π, then combining that path with the image under f of one in the link of P in D connecting incoming to outgoing rays of γ (note that such a path in $\text{Link}_D(P)$ has angular measure precisely π) would yield a nontrivial circuit in the link of $f(P)$ in Y of total angular measure $< 2\pi$. This is contrary to the assumption of non positive curvature on Y.

DEFINITION 4.7: Let $f : D \to Y$ satisfy the hypotheses of Lemma 4.7. A subpolyhedral disc $E \subset D$ is called convex if the angle sum at each boundary point of E is at most π. Note that we do not require E to be a subcomplex for given cell structure on D.

PROPOSITION 4.8. *Let $f : D \to Y$ satisfy the hypotheses of Lemma 4.7, where Y in addition is simply connected, and let E be a convex subpolyhedron of D. Then f maps E injectively onto its image.*

PROOF: Suppose $P, Q \in E$ are distinct points such that $f(P) = f(Q)$. Let γ be the geodesic segment connecting P to Q. Observe that by convexity γ lies entirely in E. By Lemma 4.6, the image of γ under f satisfies the geodesic condition except at the endpoints. Since Y is simply connected, this means that $f \circ \gamma$ bounds a geodesic monogon. But this cannot exist in non positive curvature, by the piecewise Euclidean analog of the Gauss-Bonnet theorem [GS1]. It follows that f is injective on E.

Now we return to the 2-complex $X = K(\mathcal{P})$, where $\mathcal{P} = \langle x, y \mid yx^p y^{-1} = x^q \rangle$ for positive p, q. Let Y be the $\mathbb{Z}$-cover of X. We observe that Y has a piecewise Euclidean structure of non positive curvature (although X itself does not unless $p = q$). In this structure all angles are either $\frac{\pi}{2}$ or π. A picture of this piecewise Euclidean structure can be gotten by examining Figure 5, for the special case $p = 1, q = 2$. Here the covering transformation corresponding to y is a homothety on the edges covering x, and multiplies lengths of such edges by 2. In the general case, y is a homothety on edges covering x with a scale factor of $\frac{q}{p}$. This piecewise Euclidean structure lifts to one on $\tilde{X}$ so $\tilde{X}$ becomes a piecewise Euclidean 2-complex of non positive curvature.

Observe now that each diagram $\tilde{\mathcal{D}}_n$ can be imbedded in some $\tilde{\mathcal{D}}_N$ for large enough N in such a way that the smaller diagram is contained in a convex subpolyhedron of the larger one. Observe further that all interior angle sums in all of these diagrams is exactly π (see Figure 5 for the special case $p = 1, q = 2$). It follows from Proposition 4.8 that each map $\tilde{f}_n : \tilde{\mathcal{D}}_n \to \tilde{X}$ is injective. We have already observed that this implies there is no cancellation in the chain $c(\tilde{\mathcal{D}}_n)$. Since the lengths of the boundary labels of these diagrams increase linearly with n, whereas their l_1-norms, and hence their areas, increase exponentially, this completes the geometric proof of Theorem B.

5. Proof of Theorem C

We are given $f : X \to Y$, a π_1-injective map of finite aspherical 2-complexes. We first treat the special case when $X^{(1)}$ is a subcomplex of $Y^{(1)}$ and $f|_{X^{(1)}}$ is the inclusion $X^{(1)} \subseteq Y^{(1)}$. Choose the base point in X

and let $G = \pi_1(X)$, $H = \pi_1(Y)$. We note that f lifts to a map $\tilde{f} : \tilde{X} \to \tilde{Y}$ of universal covers. In addition each 2-cell of $\tilde{X}$ determines upon applying $\tilde{f}$ a van Kampen diagram in $\tilde{Y}$ whose associated chain in $C_2(\tilde{Y})$ is well defined, since Y is aspherical. We have in fact

LEMMA 5.1. *The induced map* $\tilde{f}_2 : C_2(\tilde{X}) \to C_2(\tilde{Y})$ *is an injective homomorphism of* $\mathbb{Z}G$ *modules.*

PROOF: Consider the commutative diagram in Figure 6 below.

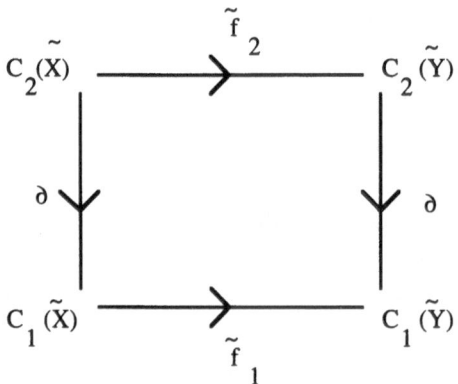

Figure 6

Observe that $\tilde{f}_1$ is a split injection since $X^{(1)} \subseteq Y^{(1)}$ and the vertical arrows are injective since X and Y are aspherical. The result follows by a diagram chase.

Hence we can consider the short exact sequence of chain complexes

(5.2) $$0 \to C_*(\tilde{X}) \xrightarrow{\tilde{f}_*} C_*(\tilde{Y}) \to Q_* \to 0$$

where Q_* is the quotient complex. Note that $C_*(\tilde{X})$ and $C_*(\tilde{Y})$ possess homology only in degree 0 where $H_0 = \mathbb{Z}$ is mapped isomorphically by $\tilde{f}_0$. It follows that $H_i(Q_*) = 0$ for all $i \geq 0$. Thus we have the short exact sequence of $\mathbb{Z}G$ modules

(5.3) $$0 \to Q_2 \to Q_1 \to Q_0 \to 0.$$

But we assumed f to be an inclusion on the 1-skeleta. This implies that Q_0 and Q_1 are free modules over $\mathbb{Z}G$, since $\mathbb{Z}H$ is free over $\mathbb{Z}G$. Thus (5.3)

splits and Q_2 is projective over $\mathbb{Z}G$. Hence $\tilde{f}_2 : C_2(\tilde{X}) \to C_2(\tilde{Y})$ is a split injection of free $\mathbb{Z}G$ modules. Let $\rho' : C_2(\tilde{Y}) \to C_2(\tilde{X})$ be a splitting, so $\rho' \cdot \tilde{f}_2 = 1$. Note that $C_2(\tilde{X})$ is finitely generated and free over $\mathbb{Z}G$, although $C_2(\tilde{Y})$ need not be finitely generated.

LEMMA 5.4. *Let $f : A \to B$ be a split monomorphism of free modules over a ring R where A is finitely generated. Let B have the free basis $\{\, b_i \mid i \in I \,\}$. Then there exists a splitting $\rho : B \to A$ for f such that $\rho(b_i) = 0$ for all but a finite number of indices $i \in I$.*

PROOF: Let $a_1, a_2, \ldots, a_n$ be a finite free basis for A and let $\rho' : B \to A$ be a splitting, so that $\rho' \cdot f = 1_A$. Since $f(a_j)$ is a finite linear combination of the b_i's, the set I_0 of indices i for which b_i has a non zero coefficient in some $f(a_j)$ is finite. We define $\rho : B \to A$ by $\rho(b_i) = \rho'(b_i)$ if $i \in I_0$ and $\rho(b_i) = 0$ if $i \notin I_0$. Then one checks that $\rho \cdot f = 1_A$.

We apply the lemma with $A = C_2(\tilde{X})$ and $B = C_2(\tilde{Y})$. The basis for $C_2(\tilde{Y})$ consists of multiples $t \cdot \tilde{\beta}$ where $\tilde{\beta}$ is a chosen lift to $\tilde{Y}$ of the 2-cell β of Y and where t describes a transversal of coset representatives for the image of G in H. We obtain a retraction $\rho : C_2(\tilde{Y}) \to C_2(\tilde{X})$, $\rho \cdot \tilde{f}_2 = 1$, with $\rho(t \cdot \tilde{\beta}) = 0$ for all but a finite number of coset representatives t. But this means that the matrix M for ρ in terms of the free bases for $C_2(\tilde{X})$ and $C_2(\tilde{Y})$ as free $\mathbb{Z}G$ modules has only a finite number of non zero entries. Thus M is bounded in l_1-norm and if we set A equal to the supremum of the l_1-norms of the matrix entries (in $\mathbb{Z}G$) of M, then we have

$$(5.5) \qquad\qquad |\rho(u)|_1 \le A\, |u|_1$$

for all $u \in C_2(\tilde{Y})$. If we let $c \in C_2(\tilde{X})$ and calculate l_1-norms, we get $|c|_1 = |\rho \cdot \tilde{f}_2(c)|_1 \le A \cdot |\tilde{f}_2(c)|_1$, where we have used (5.5) in the last inequality. In particular this holds if $c = c(\tilde{g})$, where $c(\tilde{g})$ is the 2-chain determined by the lift $\tilde{g}$ of a van Kampen diagram g for a circuit w in $X^{(1)}$; here w is assumed null homotopic in X. But this means that

$$(5.6) \qquad\qquad |w|_{1,X} \le A\, |w|_{1,Y}$$

for all such circuits w, where the subscripts X and Y refer to the l_1-norms for these respective spaces. Since (5.6) holds for all such circuits w, we deduce that

$$(5.7) \qquad\qquad \lambda_X(n) \le A\, \lambda_Y(n)$$

for all $n > 0$. This completes the proof of Theorem C in the special case.

In general, given $f : X \to Y$ a π_1-injective map of finite aspherical 2-complexes where Y has at least one 2-cell, we can modify Y by elementary 2-expansions to get the finite 2-complex Y' containing Y and a copy of $X^{(1)}$ as subcomplexes so that the map f is homotopic to a map $f' : X \to Y'$ where f' restricted to $X^{(1)}$ is the inclusion of $X^{(1)}$ in $Y'^{(1)}$. The result follows from Theorem 3.5 and the special case just considered. This completes the proof of Theorem C.

REMARK 5.8: The hypotheses of asphericity of the 2-complexes in Theorem C cannot be dropped as the following example shows. Let $G = G(\mathcal{P})$ where $\mathcal{P} = \langle x, y, z, \mid [x, y] = z, [z, x] = 1, [z, y] = 1 \rangle$; G is the 3-dimensional integral Heisenberg group. For the record, the $(2n + 1)$-dimensional integral Heisenberg group has a presentation $\langle x_1, \ldots, x_n, y_1, \ldots, y_n, c \mid$ all generators commute except $[x_i, y_i] = c, \ 1 \leq i \leq n \rangle$. Let H be the 5-dimensional Heisenberg group. According to Thurston [Th], H satisfies the quadratic isoperimetric inequality, although G does not. We shall prove this assertion about G in the next paragraph 5.9 by exhibiting a family of null homotopic circuits w_n whose area increases like a cubic function of n although the length of w_n is a linear function of n. Then in Proposition 5.12 below we shall show that the area of w_n, when considered in H, is bounded by a quadratic function of n. At the time of writing, we do not know if our arguments can be extended to show that H satisfies the quadratic isoperimetric inequality.

REMARK 5.9: Another presentation for the 3-dimensional Heisenberg group G is $\mathcal{Q} = \langle x, y, t \mid [x, y] = 1, \ txt^{-1} = xy, \ tyt^{-1} = y \rangle$, as one can easily check (in this presentation, y represents the central commutator). This presentation is Cockcroft (that is, its associated 2-complex $K(\mathcal{Q})$ is Cockcroft). To see this, note that the 2-complex $K(\mathcal{Q})$ is the 2-skeleton of a cell structure of a closed, orientable, aspherical 3-manifold M with precisely one 3-cell. The manifold M is in fact a torus bundle over the circle with monodromy the automorphism ϕ of $\mathbf{Z}^2$ given by $\phi(x) = xy, \ \phi(y) = y$. It follows from Proposition 3.7 that $\mathcal{Q}$ is Cockcroft.

Now let $w_n = [x^n, t^n x^n t^{-n}]$, where $[u, v] = uvu^{-1}v^{-1}$. In G we have that $t^n x^n t^{-n} = \phi^n(x^n) = (xy^n)^n$ commutes with x^n, so w_n is null homotopic in $X := K(\mathcal{Q})$. We can visualize a van Kampen diagram f_n for w_n as follows. The top third is a trapezoid that gives a null homotopy for $u_n := t^n x^n t^{-n} \phi^n(x^n)^{-1}$. The bottom third is the reflection of the top third through a horizontal line. The middle third represents a van Kampen diagram for the relation $[x^n, \phi^n(x^n)] = 1$ (the middle third is glued to the top and bottom thirds along arcs whose label is $\phi^n(x^n)$). Now we can calculate the chain $c(f_n)$ for f_n in $C_2(X)$ and hence its l_1-norm as follows. The top and bottom thirds of f_n are oppositely oriented, so their chains exactly cancel. It remains to calculate the chain for the middle third. There is a short cut for doing this. Commutation of x with itself doesn't count (it represents a one dimensional subset of the diagram). All we need to count are the number of commutations between x^n and the y's occurring in $\phi^n(x^n)$. The easy way to do this is to use the Grassman product, $\Lambda^2(\mathbb{Z}^2)$ equipped with its basis $x \wedge y$ (and the l_1-norm with respect to this basis for this rank 1 lattice). Calculating $[x^n] \wedge [(xy^n)^n]$ and using additive notation instead of multipicative notation, we get $nx \wedge (nx + n^2 y) = n^3 x \wedge y$. It follows from Remark 3.6 that the l_1-norm of w_n is estimated below by $|c(f_n)|_1 = n^3$.

Since $\ell(w_n) = 8n$ increases linearly with n, whereas its l_1-norm, calculated in X, increases like a cubic polynomial in n, it follows from Lemma 3.4 that the Dehn function for G is at least cubic, and hence that G is not automatic. In fact one can show easily that there is a constant $B > 0$ such that $\delta_X(n) \leq Bn^3$ for all $n > 0$. Hence G satisfies a cubic isoperimetric inequality and this cubic inequality is optimal.

Summarizing, we have the result that *the 3-dimensional Heisenberg group does not satisfy the quadratic isoperimetric inequality.*

REMARK 5.10: We shall show more generally *the split extension $G = \mathbb{Z}^2 \rtimes_\phi \mathbb{Z}$, where $\phi \in \mathrm{Aut}(\mathbb{Z}^2) \cong \mathrm{Gl}_2(\mathbb{Z})$, is automatic iff ϕ is of finite order.* The 'if' direction is easy, for if ϕ is of finite order, then G contains a free abelian group of rank 3 as a subgroup of finite index. It follows then from results of [CEHPT that G is automatic.

In the converse direction, we shall show the stronger result that *if ϕ is not of finite order, then G does not satisfy the quadratic isoperimetric inequality.* Up to replacing G by a subgroup of index 2 (this does not change the growth of the Dehn function, by Proposition 2.11), we may assume that the determinant of ϕ is 1. In this case, if $\{x, y\}$ is a basis for the normal $\mathbb{Z}^2$ subgroup, the presentation $\mathcal{P} = \langle x, y, t \mid [x, y] = 1, txt^{-1} = \phi(x), \, tyt^{-1} = \phi(y) \rangle$ is Cockcroft, since its associated 2-complex is the 2-skeleton of a cell structure on a closed, orientable, aspherical 3-manifold with exactly one 3-cell.

In this case there are two subcases, depending on whether ϕ has 1 as an eigenvalue or not. If 1 is an eigenvalue, then the basis $\{x, y\}$ may be chosen so that the matrix for ϕ with respect to this basis has the form $\begin{pmatrix} 1 & 0 \\ a & 1 \end{pmatrix}$, where $a \neq 0$. Then the argument given in Remark 5.9 applies unchanged to show that the G satisfies the cubic isoperimetric inequality and this is optimal (the sequence of words $w_n = [x^n, t^n x^n t^{-n}]$ has length linear in n but area cubic in n; the key point as before is the fact that the presentation $\mathcal{P}$ is Cockcroft).

Suppose now that $\det(\phi) = 1$, ϕ is of infinite order, and no eigenvalue of ϕ is 1. In this case it follows from the classification of elements of $Sl_2(\mathbf{Z})$ that ϕ is a hyperbolic automorphism. It has two real eigenvalues $\lambda, \frac{1}{\lambda}$ and we may assume that $|\lambda| > 1, |\frac{1}{\lambda}| < 1$. In addition we may assume, after replacing ϕ by ϕ^2 if necessary, that $\lambda > 1$. Furthermore, an elementary number theoretic argument based on the quadratic formula for λ shows that λ is irrational. Let ξ, η be (real) eigenvectors for ϕ for the eigenvalues $\lambda, \frac{1}{\lambda}$, respectively; here ϕ is considered to be in $Sl_2(\mathbf{R})$. We have $x = \alpha\xi + \gamma\eta$, $y = \beta\xi + \delta\eta$ for suitable real numbers α, β, γ and δ. This takes place in $\mathbf{R}^2 = \mathbf{R} \otimes \mathbf{Z}^2$. A calculation shows that

$$\phi^n(x) = (\alpha\delta\lambda^n - \frac{\beta\gamma}{\lambda^n})x - \alpha\gamma(\lambda^n - \frac{1}{\lambda^n})y.$$

Observe that, in this last expression, $\alpha\gamma \neq 0$, for otherwise x would be an eigenvector of ϕ, contradicting the irrationality of λ. It follows that the coefficient of y in $\phi^n(x)$, which is of course an integer, increases like λ^n; that is, it increases exponentially with n. From this point on, the argument is the same as in Remark 5.9 using the same family of words w_n. In this case one gets the stronger result that the area of w_n increases exponentially with n although the length of w_n increases linearly. Since we have analyzed all cases, it follows that G does not satisfy the quadratic isoperimetric inequalty if ϕ is of infinite order.

We shall now show that the family of words $w_n = [x^n, t^n x^n t^{-n}]$, which were introduced in Remark 5.9 to show that the cubic bound for the Dehn function of the 3-dimensional integral Heisenberg group is optimal, have a quadratic bound on area when considered in the 5-dimentional integral Heisenberg group.

Let $Y = K(\mathcal{Q})$, where

$$\mathcal{Q} = \overset{\bullet}{\langle} x, y, c, t, u \mid \text{all generators commute except } [t, x] = c, [u, y] = c \rangle.$$

The presentation $\mathcal{Q}$ is one of the 5-dimensional Heisenberg group H. We shall prove here

PROPOSITION 5.11. *There exists a constant $A > 0$ so that* $\text{Area}_Y(w_n) \leq An^2$ *for all $n > 0$.*

PROOF: It is convenient to work with yet another presentation of H. We do Tietze expansions on Q by introducing four new generators a, b, α, β with the added relations $a = xy^{-1}, b = tu, \alpha = xy, \beta = tu^{-1}$. We also adjoin the relators $[a, b], [b, t], [b, u], [\beta, \alpha], [\beta, t], [\beta, u]$, which a calculation shows are consequences of the previous relators. Denote by $\mathcal{R}$ the resulting presentation for H and let $Z = K(\mathcal{R})$. Since the area function changes at most by a linear change of variable plus a linear function under Tietze transformations by Theorem 2.7, to prove Proposition 5.11 it suffices to prove there is a positive constant B so that $\text{Area}_Z(w_n) \leq Bn^2$.

DEFINITION: We call the relators of $\mathcal{R}$ which say two generators commute or which define a, b, α and β in terms of the previous generators Type I relators. The two remaining relators, $[t, x] = c$, and $[u, y] = c$, we call Type II relators. If w, w' are words in the generators of $\mathcal{R}$, we write $w \sim w'$ if w' can be obtained from w by applying a sequence of at most $M(w, w')^2$ of the conjugates of the Type I relators, where $M(w, w')$ is the larger of the lengths of the words w, w'. Geometrically this means that the word $w'w^{-1}$ bounds a van Kampen diagram in $\mathcal{R}'$ of area at most $M(w, w')^2$, where $\mathcal{R}'$ is the presentation with the generators of $\mathcal{R}$ and only the Type I relators.

We need the following

LEMMA 5.12. *There is a positive constant C with the following property: if $w(x, t)$ is a word of length n in the generators x, x^{-1}, t, t^{-1} whose exponent sums in $x^{\pm}$ and in $t^{\pm}$ are both 0, then*

$$\text{Area}_Z(w(x, t)w(x^{-1}, t)) \leq Cn^2.$$

PROOF OF LEMMA 5.12: The hypothesis on the exponent sums means w represents 1 in the free abelian group of rank 2 with basis x, t. Using only the relators of Type I, we see that

$$w(x^{-1}, t) \sim w(x^{-1}y^{-1}, tu^{-1}u) \sim w(\alpha^{-1}, \beta u) \sim w(\alpha^{-1}, u) \sim w(y^{-1}, u).$$

Hence we have

$$w(x, t)w(x^{-1}, t) \sim w(x, t)w(y^{-1}, u) \sim w(xy^{-1}, tu) \sim w(a, b) \sim 1.$$

Since only a bounded number of "$\sim$'s" appear, it follows that there exists $C > 0$ such that $\text{Area}_Z(w(x,t)w(x^{-1},t)) \leq Cn^2$. This completes the proof of the lemma.

Returning to the proof of Proposition 5.11, we specialize to the words $w_n = [x^n, t^n x^n t^{-n}]$. Note that w_n is formally cyclically conjugate to the word $t^n x^n t^{-n} x^{-n} t^n x^{-n} t^{-n} x^n$ and this latter word is of the form $u_n(x,t)u_n(x^{-1},t)$, where $u_n(x,t) = [t^n, x^n]$. These words are of length a linear function of n, so it follows from Lemma 5.12 above that there exists a constant $B > 0$ so that $\text{Area}_Z(w_n) \leq Bn^2$. This completes the proof of Proposition 5.11.

REMARK 5.13: By virtue of Theorem C, there is interest in producing examples of groups G of geometric dimension 2 whose Dehn function is equivalent to its l_1-norm function; here G is said to be of geometric dimension 2 if $G = \pi_1(X)$, where X is a finite aspherical 2-complex. For example, suppose $G < H$, where G and H are both groups of geometric dimension 2 and suppose that H is hyperbolic. If the Dehn function for G is equivalent to its l_1- norm function, then G is also hyperbolic. To see this, observe that by Theorem C, λ_G is dominated by λ_H, which is in turn dominated by δ_H. By assumption, λ_G is equivalent to δ_G. Since H is hyperbolic, it follows that δ_H is linearly bounded. Consequently δ_G is also linearly bounded, and hence G is hyperbolic. In the next paragraph, we shall give one example where λ_G is equivalent to δ_G.

EXAMPLE 5.14: We shall show that if $X = K\langle x, y \mid [x,y] = 1\rangle$, then $\lambda_X = \delta_X$. Observe first that $\tilde{X} = \mathbf{R}^2$, where $\tilde{X}$ is the universal cover of X. If w is an edge circuit in $X^{(1)}$ which is null homotopic in X and such that no subpath of w is null homotopic, then w lifts to a simple closed circuit $\tilde{w}$ in $\mathbf{R}^2$. By the Jordan-Schoenfliess theorem, $\tilde{w}$ bounds a disc, which gives an *injective* disc diagram $f : D^2 \to \tilde{X}$ bounded by $\tilde{w}$. Hence $|c(\tilde{f})|_1 = \text{Area}_X(w)$. From this observation, it is not difficult to complete the proof that $\lambda_X = \delta_X$.

QUESTION: It seems reasonable to ask whether $\lambda_X = \delta_X$ for every finite connected 2-complex X admitting a piecewise Euclidean structure of non positive curvature (see 4.5). We can also ask whether $\lambda_X \sim \delta_X$ for every finite connected 2-complex X.

6. Examples

Early on in the discussions leading to [BGSS] the authors asked whether an amalgam of finitely generated free groups over a subgroup of finite index in each was automatic. We observed

PROPOSITION 6.1. *Let F be a finitely generated free group and let N be a normal subgroup of finite index. Then the double $D = F \underset{N}{\star} F$ of F along N is automatic.*

PROOF: If $G = F/N$ one has a homomorphism $\phi : G \to \text{Out}(N)$ with $\phi(aN)$ induced by conjugation by a on N for $a \in F$. If we consider the short exact sequence of groups

(6.2) $$1 \to N \to D \to G \star G \to 1,$$

then we get an induced homomorphism $\phi_1 : G \star G \to \text{Out}(N)$ with $\text{Im}\phi_1 = \text{Im}\phi$. In particular $\text{Im}\phi_1$ is a finite subgroup of $\text{Out}(N)$, so the kernel K_1 of ϕ_1 is a normal subgroup of finite index in $G \star G$. But the kernel K_2 of the canonical homomorphism $G \star G \to G \times G$ is of finite index and free. Let $K_3 = K_1 \cap K_2$. It follows that K_3 is finitely generated free and of finite index in $G \star G$ and $\phi_1|K_3$ is trivial. This implies that the extension (6.2) pulled back to K_3 is the direct product of K_3 with N. That is, D possesses a subgroup of finite index which is isomorphic to $N \times K_3$, the direct product of two finitely generated free groups. Thus $N \times K_3$ is automatic [CEHPT]. Since D has a subgroup of finite index which is automatic, D is also automatic.

However in general an amalgam of two finitely generated free groups over a subgroup of finite index in each is not automatic. The following construction is due to Stallings. Let F_1 be freely generated by elements a and s and let F_2 be freely generated by b and t. Let $A_1 = \text{gp}\langle a, sa\bar{s}, s^2as^{-2}, s^3 \rangle < F_1$ and let $A_2 = \text{gp}\langle b, tb^2t^{-1}, t^2bt^{-1}, t^3 \rangle < F_2$. Observe that $(F_i : A_i) = 3$ for $i = 1, 2$. Thus we can form $H = F_1 \underset{A_1 \cong A_2}{\star} F_2$ where we identify $a = b, sas^{-1} = tb^2t^{-1}, s^2as^{-2} = t^2bt^{-1}$, and $s^3 = t^3$. Then $t^{-1}sas^{-1}t = b^2 = a^2$, so $a^{(t^{-1}s)} = a^2$. One sees easily that the subgroup $\text{gp}\langle a, t^{-1}s \rangle$ of H is in fact isomorphic to $G\langle x, y \mid yx\bar{y} = x^2 \rangle$. But $H = \pi_1(Y)$ with Y a finite aspherical 2-complex. It follows from Theorem A that H is not automatic.

EXAMPLE 6.3: Let $\mathcal{P}_r = \langle x_0, x_1, \ldots, x_r \mid x_i^{x_{i+1}} = x_i^2; 0 \le i \le r - 1 \rangle$. The Dehn function for $\mathcal{P}_r$ grows no slower than the function E_r; here $E_0(n) = n$

and $E_{r+1}(n) = 2^{E_r(n)}$ for all $n > 0, r > 0$. Let us sketch why this is so (the case $r = 1$ is contained in the proof of Theorem A). We consider first $\mathcal{P}_1$ and the words $w_{1,n} = [x_0^{x_1^n}, x_0]$. The methods of §4 show that the minimal van Kampen diagram for $w_{1,n}$ has $2(2^n - 1) \geq 2^n$ faces (see Figure 5) and that this number of faces is the same as the l_1-norm. Now specialize to $w_{1,2^n} = [x_0^{x_1^{2^n}}, x_0]$. We replace $x_1^{2^n}$ by $x_1^{(x_2^n)}$ to get $w_{2,n} = [x_0^{(x_1^{(x_2^n)})}, x_0]$. One has $l(w_{2,n}) = 2^3(n + 1)$ and the minimal van Kampen diagram for it has at least $E_2(n) = 2^{2^n}$ faces and the number of faces is the l_1-norm. At the next stage we specialize to $w_{2,2^n}$ and replace $x_2^{2^n}$ by $x_2^{(x_3^n)}$. In general we have words $w_{k,n}$ with $l(w_{k,n}) = 2^{k+1}(n + 1)$ but the minimal van Kampen diagram for $w_{k,n}$ has at least $E_k(n)$ faces and the number of faces is the l_1-norm. If we attempt to draw these minimal van Kampen diagrams the appearance is similar to the approximations to the Koch snowflake curve.

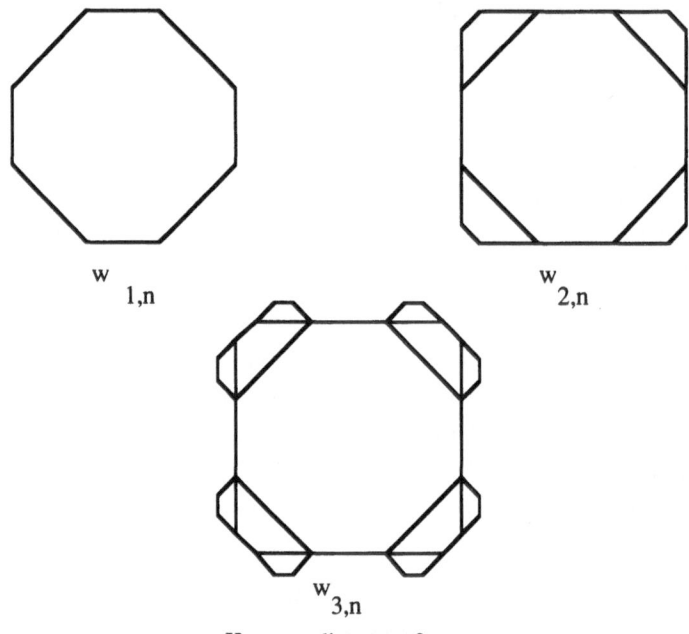

$$w_{1,n} \qquad w_{2,n}$$

$$w_{3,n}$$

van Kampen diagrams for $w_{k,n}$

Figure 7

Next consider the 1-relator presentation $\mathcal{P} = \langle x, y \mid x^{x^y} = x^2 \rangle$. The group $G(\mathcal{P})$ maps onto $\mathbb{Z}$ by $x \mapsto 0, y \mapsto 1$. The kernel K can be calculated by the Reidemeister-Schreier method. It has the presentation $\mathcal{Q} = \langle\, x_i;\, i \in \mathbb{Z} \mid x_i^{x_{i+1}} = x_i^2$ for all $i \in \mathbb{Z} \,\rangle$. In particular $\mathcal{P}_r$ is a subpresentation of $\mathcal{Q}$ for all $r \geq 1$ and the group imbeds as well.

By Theorem C, since $K(\mathcal{P})$ is aspherical, the l_1-norm function for $\mathcal{P}$ grows at least as fast as that for any of the presentations $\mathcal{P}_r$. Thus the Dehn function for $\mathcal{P}$ grows faster than the function E_r for all r. With more care one can actually show that $\delta_{\mathcal{P}}$ grows at least as fast as Δ, where $\Delta(n) = E_n(n)$. This is indeed a fast rate of growth and is our record so far for 1-relator groups.

PROPOSITION 6.4. Let $\mathcal{R} = \langle\, x, y, z, \mid x^{x^y} = x^2 \,\rangle$. Then $G = G(\mathcal{R})$ is not an amalgam of finitely generated free groups amalgamating a finitely generated subgroup.

PROOF: The group $G(\mathcal{R})$ admits the group $G(\mathcal{P})$ above as a retract, so the Dehn function for $\mathcal{R}$ grows at least as fast as that of $\mathcal{P}$. In particular, since $\delta_{\mathcal{P}}$ grows faster that E_2, where $E_2(n) = 2^{2^n}$, it follows that $\delta_{\mathcal{P}}$ grows faster than any simple exponential. But an amalgam of finitely generated free groups amalgamating a finitely generated subgroup is asynchronously automatic by [BGSS], so its Dehn function is bounded by a simple exponential. Hence $\mathcal{R}$ is not an amalgam of such groups.

REMARK 6.5: This result answers negatively a question raised by G. Baumslag. Baumslag and Shalen [BS] proved that a finitely presented group admitting a finite presentation of deficiency at least two is a proper amalgam of finitely generated subgroups over a finitely generated amalgamated subgroup. Baumslag asked whether in the case of a 1-relator group of deficiency at least two these subgroups could all be taken to be finitely generated and free. The example $\mathcal{R}$ above shows that this is not the case.

REMARK 6.6: It can be shown that the group $G(\mathcal{P}_2)$ above admits a finite complete rewriting system [Br]. Thus a finitely presented group with a presentation that admits a finite complete rewriting system need not be asynchronously automatic. The group $G(\mathcal{P}_1)$ provides a simpler example of one which admits a finite complete rewriting system but is not synchronously automatic.

We finish with some open problems. We have not yet been able to decide whether the Baumslag-Solitar group $B_{1,2}$ is isomorphic to a subgroup of an automatic group. Indeed, in an earlier version of this article, we stated that we knew no restriction on the finitely presented subgroups H of general automatic groups other than that H must have a solvable word problem. Since then, Paul Schupp informed me of the fact that the theory of time-complexity [HU] puts restriction on such subgroups, based on the result that the word problem for automatic groups is solvable in $O(n^2)$-time [CEHPT].

In general, the diagrammatic algorithm we have considered in this paper is very inefficient. Its advantage is its universality, since a finitely presented group has a solvable word problem if and only if its Dehn function is recursive (Proposition 2.2 above). For example, $\mathbf{Z}^n$ satisfies the quadratic isoperimetric inequality but not the linear isoperimetric inequality if $n \geq 2$. However there is a linear time algorithm for the word problem of the presentation $\langle x_i; 1 \leq i \leq n \mid [x_i, x_j]; 1 \leq i, j \leq n \rangle$. Namely, a word represents 1 if and only if its exponent sum in each variable is zero.

More generally we do not know whether $B_{1,2}$ can be isomorphic to a subgroup of a finitely presented group satisfying the quadratic isoperimetric inequality. By Thurston's results [Th] this last class of groups is very large, including $\mathrm{Sl}_n(\mathbf{Z})$ for all $n \geq 4$ and many nilpotent groups, like the $(2n+1)$-dimensional integral Heisenberg group for all $n \geq 2$. A characterization of finitely presented groups satisfying the quadratic isoperimetric inequality is lacking unlike the situation for the linear isoperimetric inequality: a finite presentation $\mathcal{P}$ satisfies $\delta_{\mathcal{P}}(n) \leq An$ for all $n > 0$ iff $G(\mathcal{P})$ is hyperbolic [Gr].

Finally note the following immediate consequence of Theorem C: *Let* $X \overset{j}{\subseteq} Y \subseteq Z$ *be inclusions of finite connected subcomplexes X and Y of the aspherical 2-complex Z. Assume (i) j is π_1-injective, (ii) λ_X grows faster than any polynomial, and (iii) δ_Y is polynomially bounded. Then Y is not aspherical.* If the hypotheses (i)–(iii) could be realized in an example, this would give a negative answer to a question of J.H.C. Whitehead's, whether a subcomplex of an aspherical 2-complex is aspherical.

In fact, we remark that there are natural candidates among which to look for examples; namely, one should look among the labelled oriented trees, or LOT's, of Howie's [Ho]. A LOT is a pair (T, λ) where T is a tree (in Serre's sense [Se], so that geometric edges correspond to pairs of oriented edges $e, \bar{e}$ of T) together with a function $\lambda : E \to F(V)$ so that $\lambda(\bar{e}) = \lambda(e)^{-1}$. Here E is the set of edges and $F(V)$ is the free group freely generated by the set V of vertices of T. The presentation $\mathcal{P}(T, \lambda)$ associated to the LOT (T, λ) is given by $\langle V \mid (\partial_1 e)^{\lambda(e)} = \partial_0 e; e \in \mathcal{O} \rangle$, where $\mathcal{O}$ is an *orientation* on E, so $\#(\mathcal{O} \cap \{e, \bar{e}\}) = 1$ for all $e \in E$.

One verifies that the LOT (T', λ') illustrated below in Figure 8 has associated presentation $\mathcal{P}' = \mathcal{P}(T', \lambda')$ with group $G(\mathcal{P}') \cong G\langle x, y \mid yx\bar{y} = x^2 \rangle$.

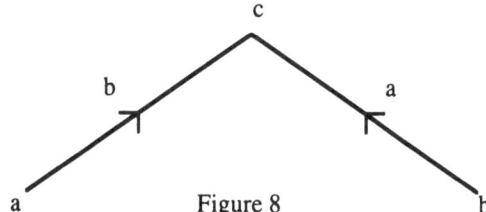

Figure 8

Thus, if this LOT (T', λ') is a sub-LOT of the LOT (T, λ), then $X :=$ $K(\mathcal{P}(T', \lambda'))$ satisfies hypothesis (ii) above. In addition, it is easy to see, by an Euler characteristic argument, that if $\mathcal{Q}$ is obtained from $\mathcal{P}(T, \lambda)$ by adjoining any one of the vertices of T as a new relator, then $Z := K(\mathcal{Q})$ is contractible, since $\pi_1(Z)$ is trivial. One takes $Y := K(\mathcal{P}(T, \lambda))$. We do not know whether (i) and (iii) can both be satisfied in an example.

REFERENCES

[A] A. D. Aleksandrov, *Odna teorema o treugolnikah v metricheskom prostranstve i nekotorye ee prilozheniya*, Trudy Mat. Inst. Im. V. A. Steklova **38** (1951), 5–23.

[BGSS] G. Baumslag, S. M. Gersten, M. Shapiro, and H. Short, *Automatic groups and amalgams*, –a summary appears in this volume.

[Br] K. S. Brown, *The geometry of rewriting systems: a proof of the Anick-Groves-Squier theorem*, preprint.

[BS] G. Baumslag and P. B. Shalen, *Amalgamated products and finitely presented groups*, this volume.

[CDP] M. Coornaert, T. Delzant, and A. Papadopoulos, *Notes sur les groupes hyperboliques de Gromov*, Publication de I.R.M.A., Strasbourg.

[CEHPT] J. W. Cannon, D. B. A. Epstein, D. F. Holt, M. S. Paterson, and W. P. Thurston, *Word processing and group theory*, preprint.

[G] S. M. Gersten, *Reducible diagrams and equations over groups*, in "Essays in Group Theory," M.S.R.I. series Vol.8, S. M. Gersten, editor, Springer-Verlag, 1987.

[GH] W. Ballmann, E. Ghys, A. Haefliger, P. de la Harpe, E. Salem, R. Strebel, et M. Troyanov, *Sur les groupes hyperboliques d'après Mikhael Gromov*.

[Gr] M. Gromov, *Hyperbolic groups*, in "Essays in Group Theory," M.S.R.I. series Vol. 8, S. M. Gersten, editor, Springer-Verlag, 1987.

[GS1] S. M. Gersten and H. Short, *Small cancellation theory and automatic groups*, Invent. Math. **102** (1990), 305–334.

[GS2] S. M. Gersten and H. Short, *Small cancellation theory and automatic groups, part II*, Invent. Math. (to appear).

[Ho] J. Howie, *On the asphericity of ribbon disc complements*, Trans. Amer. Math. Soc. **289** (1985), 281-302.

[HU] J. Hopcroft and J. Ullman, "Introduction to Automata Theory, Languages, and Computation," Addison-Wesley, 1979.

[LS] R. C. Lyndon and P. E. Schupp, "Combinatorial Group Theory," Springer-Verlag, 1977.

[Re] Yu. G. Reshetnyak, *Ob odnom special'nom otobrazhenii konusa na mnogogrannik*, in Russian, Mat. Sbornik **53(95)** (1961).

[Se] J.-P. Serre, "Trees," Springer-Verlag, 1980.

[Th] W. P. Thurston, oral communication.

Mathematics Department, University of Utah, Salt Lake City, UT 84112

Problems on Automatic Groups

S. M. GERSTEN*

This is a report on the organizational meeting of a seminar on automatic groups which took place at MSRI on 17 January 1989. Although the intention was merely to establish a time and a first speaker, Bill Thurston soon took the floor and one after another of the participants proposed questions on automatic groups, none of which could be answered at that time. It seemed worthwhile to record those questions asked as a guide to research into automatic groups. Participants proposing questions were Thurston, G. Baumslag, M. Shapiro, H. Short, and the author. We have not attempted to record who proposed what problem. After a draft of this note was sent to David Epstein, he proposed some problems he thought ought to be included. Geoff Mess also contributed suggestions after the first draft was circulated. He remarked that a notion of automatic groups was considered by Russian writers in the 1970's; it is important to point out that this earlier notion has nothing to do with the notion introduced in [CEHPT], which is the subject of this problem set.

An *automatic group* G has a finite finite set of semigroup generators S and a regular language $\mathcal{L} \subseteq S^*$, where S^* is the free monoid on S, so that every element of G is represented by at least one word in $\mathcal{L}$ and such that the question whether or not two words in $\mathcal{L}$ represent elements of G which are at most a unit apart in the Cayley graph can be decided by a finite state automaton. There is a companion notion of asynchronous automatic group which is based on two tape automata [RS] which read their tapes out of synchronization. In the two tape automaton which checks whether two words represent group elements differing by right multiplication by a generator, the machine is allowed to read letters from one string or the other (but not both simultaneously), according to programmed instructions in the machine, until one runs out of letters; then the pair of words is accepted or rejected according to the state of the machine. In the automatic structure, by way of contrast, a pair of letters, one from each string, is read

*This paper is based on research partly supported by the National Science Foundation grant Nos. DMS 850-5550 and DMS 860-1376

simultaneously (the shorter of the two words having been padded by some neutral letter to make the strings of the same length). The foundation paper for automatic groups is [CEHPT] although the idea is implicit in Cannon's paper [Ca] and germs of the idea can be found in [Gr]. Also see [BGSS] in this volume for a more leisurely discussion of these notions. We have collected a bibliography of papers of interest, including several on Gromov's word hyperbolic groups.

PROBLEM 1. *Does an automatic group have a solvable conjugacy problem?*

It is shown in [BGSS] using results of C. F. Miller III that both the conjugacy problems and isomorphism problems have negative solutions for asynchronous automatic groups. The situation for automatic groups is open.

The automatic structure $(S, \mathcal{L})$ on the group G is called *symmetric* if $w \in \mathcal{L}$ iff $w^{-1} \in \mathcal{L}$.

PROBLEM 2. *Does $F_2 \times F_2$ have a symmetric automatic structure?*

Here F_2 is the free group of rank 2.

PROBLEM 3. *Let G be a discrete cocompact subgroup of* Isom $(\mathbf{H}^2 \times \mathbf{H}^2)$, *the group of isometries of the the Cartesian square of the hyperbolic plane. Suppose that G is not virtually split. Is it the case that G is not automatic?*

The group G above is called *virtually split* if there exists a subgroup $G_0 \leq G$ of finite index so that $G_0 = G_1 \times G_2$, where $G_i \in \mathrm{Isom}(\mathbf{H}^2)$, for $i = 1, 2$.

Perhaps the most powerful techniques for showing that a group does not possess an automatic structure are based on the isoperimetric inequalities due to Thurston (unpublished). The simplest of these states that for a finite presentation of the automatic group G there is a positive constant A so that every free word in the generators w which represents the identity element of G bounds a singular disc diagram with at most $A \cdot \mathrm{length}(w)^2$ faces; this is the so called *quadratic isoperimetric inequality*. For more details consult [BGSS]. Thurston asserted that $\mathrm{Sl}_3(\mathbf{Z})$ and the three dimensional integral Heisenberg group do not satisfy the quadratic isoperimetric inequality, whereas $\mathrm{Sl}_{n+2}(\mathbf{Z})$ and the $2n+1$ dimensional integral Heisenberg groups satisfy the quadratic isoperimetric inequality for $n \geq 2$. No details for these results are available at present and it is hoped that they will be provided before long. Thurston also raised the question whether a finitely generated nilpotent group which is *not* abelian by finite must necessarily fail to satisfy one of his higher dimensional isoperimetric inequalities.[1]

Little is known about the subgroups of an automatic group G other than the fact that a finitely generated subgroup has a solvable word problem. However it is known that a finitely generated subgroup of G need not be finitely presented; if one takes the kernel N of the homomorphism from $F_2 \times F_2$ to $\mathbb{Z}$ that takes each member of a free basis for each factor to a generator of $\mathbb{Z}$, then N is finitely generated but not finitely presented (in fact, $H_2(N)$ is not finitely generated as an abelian group). Here $F_2 \times F_2$ is automatic as the product of two automatic groups. More surprising is the observation due to G. Mess (unpublished) that a finitely presented subgroup of an automatic group need not be automatic. Mess takes the kernel M of the homomorphism $F_2 \times F_2 \times F_2$ to $\mathbb{Z}$ which takes each member of a free basis of each factor to a generator for $\mathbb{Z}$ and he verifies that M is finitely presented but $H_3(M)$ is not finitely generated. It is not known whether the group M satisfies the quadratic isoperimetric inequality. The next two problems relate to the subgroup structure of automatic groups.

PROBLEM 4. *Is there an automatic group G with a nilpotent subgroup N where N is not abelian by finite?*

By a result of [CEHPT] a nilpotent group is not even asynchronously automatic unless it is abelian by finite.

PROBLEM 5. *Can the group of the presentation $\langle x, y \mid yxy^{-1} = x^2 \rangle$ be isomorphic to a subgroup of an automatic group?*

The result of [Ge] shows that this group is not a subgroup of an automatic group of geometric dimension 2; in particular it is not automatic.

PROBLEM 6. *Is a retract of an automatic group automatic?*

It is not even known whether G is automatic if $G \times H$ is automatic. However M. Shapiro has shown that G is automatic if $G \star H$ is automatic [BGSS]; the proof is non trivial. It is perhaps worth pointing out that a retract H of a group G satisfies the same (or better) isoperimetric inequality

---

[1]Thurston also conjectured an improvement of his inequalities in analogy with those satisfied by a compact Riemannian manifold M of non positive curvature with convex boundary: if $\tilde{M}$ is the universal cover of M then there exists a constant $A > 0$ such that every n-cycle c_n in $\tilde{M}$ bounds an $(n + 1)$-chain b_{n+1} with $\text{Volume}(b_{n+1}) \leq A \cdot \text{Volume}(c_n)^{\frac{n+1}{n}}$.

as G itself. The difficulty in Problem 6 consists in producing the regular language for the subgroup.

PROBLEM 7. *Let G be an automatic group and let $x \in G$ be an element of infinite order. Is the map of the set of integers into the Cayley graph of G given by $n \mapsto x^n$ a quasi isometry?*

The map indicated is a quasi isometry if there exist positive constants λ and ϵ so that

$$\frac{1}{\lambda}|m - n| - \epsilon \le d(x^m, x^n) \le \lambda|m - n| + \epsilon$$

for all $m, n \in \mathbf{Z}$. Here d is the distance function in the Cayley graph. The corresponding assertion is true for word hyperbolic groups [BGSS].

PROBLEM 8. *Is the center of an automatic group finitely generated? Is there a bound on the order of the torsion elements of the center? Is G/Z automatic if G is automatic and Z is the center of G?*

PROBLEM 9. *Is there a Tits alternative valid for automatic groups? Namely, is every subgroup of an automatic group either small, in the sense of having an abelian normal subgroup of finite index, or large, in the sense of containing F_2?*

PROBLEM 10. *Let $\phi \in \mathrm{Out}(F)$, where F is a finitely generated free group, and let $G = F \rtimes_\phi \mathbf{Z}$. Is G automatic?*

It has also been conjectured that $F \rtimes_\phi \mathbf{Z}$ is word hyperbolic if it contains no subgroup isomorphic to $\mathbf{Z} \oplus \mathbf{Z}$ [GSt].

PROBLEM 11. *Which 1-relator groups are asynchronously automatic?*

Not all 1-relator groups are asynchronously automatic. The group of the presentation $\langle x, y \mid x^{x^y} = x^2 \rangle$ is not asynchronously automatic [Ge]; here $x^y = yxy^{-1}$. In this connection one should remark that 1-relator groups with torsion are word hyperbolic. This is observed in [BGSS]; it is a straightforward consequence of the spelling theorem of B. Newman [LS]. A related problem of interest is to determine which 1-relator groups are word hyperbolic and which are automatic. It may be premature to speculate, but one suggestion is that the hyperbolic 1-relator groups are precisely those which do not contain a subgroup isomorphic to any Baumslag-Solitar group $B_{k,l} = \langle x, y \mid yx^k y^{-1} = x^l \rangle$, $k \cdot l \ne 0$, and the automatic 1-relator groups are those which do not contain an isomorphic copy of $B_{k,l}$ with the further condition that $|k| \neq |l|$. The fact that these conditions are necessary lends some credibility to the conjecture.

PROBLEM 12. *Is there a finitely generated group whose Cayley graph Γ is combable but which is not automatic?*

Here Γ is called *combable* if there is a Lipschitz contraction of the vertex set of Γ to be base point. Combability implies the isoperimetric inequalities of Thurston's and the FP_∞ property [Al]. A group G is said to be of type FP_∞ if the trivial G-module $\mathbf{Z}$ possesses a free resolution by finitely generated free $\mathbf{Z}[G]$ modules. If the group G is known in addition to admit a finite presentation (as asynchronously automatic groups do), then this condition is equivalent to the existence of an Eilenberg-MacLane space $K(G,1)$ whose n-skeleton is finite for all numbers $n \geq 0$. What is lacking for the automatic structure in Problem 12 is the regular language.

PROBLEM 13. *Is a cocompact discrete subgroup of $\mathrm{SL}_3(\mathbf{R})$ automatic?*

The lattice $\mathrm{Sl}_3(\mathbf{Z})$ is of cofinite volume in $\mathrm{Sl}_3(\mathbf{R})$ and is not automatic by Thurston's result. There is a p-adic analog to this question which has been settled. Let Γ be a cocompact discrete group of isometries of the Bruhat-Tits building of $\mathrm{Sl}_3(k)$ where k is a non discrete ultrametric local field of characteristic zero (for example, the p-adic completion of the rationals). Then Γ is automatic [GSh].

PROBLEM 14. *Let $G = A \underset{C}{\star} B$ where A and B are automatic and C is infinite cyclic. Is G automatic?*

There is an analogous result of Gromov's for word hyperbolic groups. If A and B are torsion free and hyperbolic and C is a *maximal* cyclic subgroup in both, then G is hyperbolic [Gr]. M. Shapiro has shown that G is automatic if A and B are hyperbolic and C is cyclic [BGSS].

PROBLEM 15. *If G is an automatic group, does there exist an automatic structure whose regular language $\mathcal{L}$ is prefix closed and such that each element of G has precisely one representative in $\mathcal{L}$?*

It is known that one can achieve prefix closure and one can achieve unique representatives [CEHPT], but it is not known whether one can achieve both simultaneously.

PROBLEM 16. *Is an automatic group necessarily residually finite? Is it even the case that an automatic group has at least one nontrivial finite quotient group?*

It is even unknown whether word hyperbolic groups are residually finite. We recall that a group is called residually finite if the intersection of its normal subgroups of finite index is trivial.

PROBLEM 17. *Is an asynchronously automatic group of type FP_∞?*

Synchronous automatic groups are known to be of type FP_∞ [Al].

REMARK (added January, 1990): After this problem set was written and circulated, Hamish Short and the author worked out a theory of *biautomatic groups* [GSh3]. An automatic structure $\mathcal{L} \subseteq S^*$ for the group G is said to be biautomatic if the question whether two elements of $\mathcal{L}$ represent group elements that differ by left multiplication by a generator in S can be decided by a finite state automaton. Several problems in this set which are open for automatic groups have been solved for biautomatic groups. For example, biautomatic groups have a solvable conjugacy problem [GSh2], polycyclic subgroups of biautomatic groups are abelian by finite, the group $\langle x, y \mid yxy^{-1} = x^2 \rangle$ cannot be isomorphic to a subgroup of a biautomatic group, Problem 7 above has an affirmative solution for biautomatic groups, and the center of a biautomatic group is always finitely generated [GSh3]. At the time of writing, there are no examples known to distinguish the classes of automatic groups, biautomatic groups, and combable groups.

REFERENCES

[Al] J. Alonso, *Combings of Groups*, MSRI preprint 04623-89,1989.

[ABC] J. Alonso, T. Brady, D. Cooper, T. Delzant, V. Ferlini, M. Lustig, M. Mihalik, M. Shapiro, and H. Short, *Notes on negatively curved groups*, MSRI preprint 1989.

[BGSS] G. Baumslag, S.M. Gersten, M. Shapiro, and H. Short, *Automatic groups and amalgams*, –a summary appears in this volume.

[Bo] B. Bowditch, *Notes on Gromov's hyperbolicity criterion for path-metric spaces*, preprint, University of Warwick, 1989.

[Ca] J.W. Cannon, *The combinatorial structure of cocompact discrete hyperbolic groups*, Geometriae Dedicata **16** (1984), 123–148.

[Ca2] J.W. Cannon, *Negatively curved spaces and groups*, Prelilminary lecture notes from the 'Topical Meeting on Hyperbolic Geometry and Ergodic Theory', Trieste, April 1989.

[Co] D. Cooper, *Automorphisms of negatively curved groups*, preprint, University of California at Santa Barbara, 1989.

[CDP] M. Coornaert, T. Delzant, and A. Papadopoulos, *Notes sur les groupes hyperboliques de Gromov*, IRMA Strasbourg, 1989.

[CEHPT] J.W. Cannon, D.B.A. Epstein, D.F. Holt, M.S. Paterson and W.P. Thurston, *Word processing and group theory*, preprint, University of Warwick, obtainable by writing D. B. A. Epstein.

[Ep] D.B.A. Epstein, *Computers, groups and hyperbolic geometry*, Astérisque **163–164** (1988), 9–29.

[Ge] S.M. Gersten, *Dehn functions and l_1-norms of finite presentations*, these proceedings.

[GH] W. Ballmann, E. Ghys, A. Haefliger, P. de la Harpe, E. Salem, R. Strebel, and M. Troyanov, "Sur les groupes hyperboliques d'après Mikhael Gromov," edited by E. Ghys and P. de la Harpe, Birkhäuser, Progress in Mathematics Series, 1990.

[Gi] R. H. Gilman, *Groups with a rational cross-section*, in "Combinatorial Group Theory and Topology," editors S. M. Gersten and John Stallings, Annals of Math. Study 111, Princeton Univ. Press, 1986, pp. 175–183.

[Gr] M. Gromov, *Hyperbolic groups*, in "Essays in Group Theory," MSRI series Vol. 8, S.M. Gersten, editor, Springer-Verlag, 1987.

[GSh] S. M. Gersten and H. Short, *Small cancellation theory and automatic groups*, Invent. Math. **102** (1990), 305–334.

[GSh2] S. M. Gersten and H. Short, *Small cancellation theory and automatic groups, part II*, Invent. Math. (to appear).

[GSh3] S. M. Gersten and H. Short, *Rational subgroups of biautomatic groups*, MSRI preprint #07923-89, 1989, to appear in Annals of Math..

[GSt] S.M. Gersten and John Stallings, *A note on irreducible outer automorphisms of a free group*, preprint 1989.

[HU] J. E. Hopcroft and J. D. Ullman, "Introduction to Automata Theory, Languages, and Computation," Addison-Wesley, 1979.

[LS] R.C. Lyndon and P.E. Schupp, "Combinatorial Group Theory," Springer-Verlag, 1977.

[Pa] F. Paulin, *Points fixes d'automorphismes de groups hyperboliques*, Ann. Inst. Fourier, Grenoble **39,3** (1989), 651–662.

[RS] M. Rabin and D. Scott, *Finite automata and their decision problems*, IBM Jour. Res. **3:2** (1959), 115–125.

[Sh] H. Short, *Regular subgroups of automatic groups*, MSRI preprint #07723-89, 1989.

[Th] W. P. Thurston, *Finite state algorithms for the braid groups*, preprint, Princeton University, February 1989.

Mathematics Department, University of Utah, Salt Lake City, UT, 84112